FLORA OF TROPICAL EAST AFRICA

MORACEAE*

C.C. Berg**
(University of Bergen)

Trees, shrubs or herbs, dioecious or monoecious; sap milky, sometimes watery (but not turning black). Leaves in spirals or distichous, sometimes subopposite or subverticillate, entire or sometimes pinnately or palmately incised, stipulate. Inflorescences bisexual or unisexual, spicate, globose-, clavate- or discoid-capitate, urceolate or uniflorous. Staminate flowers with 2–6 tepals or perianth lacking; stamens 1–4. Pistillate flowers with 2–6 tepals or perianth lacking; pistil 1; ovary free or adnate to the perianth; stigmas 1–2; ovule 1, attached apically. Fruit achene-like, drupaceous (dehiscent or not), or forming a drupaceous whole with the fleshy perianth or with the fleshy receptacle as well. Seed large without endosperm or small with endosperm; embryo various.

A family with ± 50 genera and some 1100–1150 species, the majority tropical; ± 625 species in Asia and Australasia, ± 300 species in the Neotropics and ± 200 species in tropical Africa.

Apart from the following, introduced species of *Morus* and *Ficus* are mentioned under those genera.

Artocarpus heterophyllus Lam.; Jack Fruit; native of tropical Asia. Tree; lamina ± 10–20 × 6–10 cm., entire; inflorescences on the main branches and the trunk. Grow in higher rainfall areas in the coastal region, in the Usambara Mts. of Tanzania, and also in Uganda; sometimes grown in the midlands of Kenya as an ornamental but not for fruit.

Artocarpus altilis (Parkinson) Fosberg (*A. communis* J.R. & G. Forster); Bread Fruit; native of tropical Asia. Tree; lamina mostly ± 30–50 × 15–20 cm., pinnately incised; inflorescences in the leaf-axils. Grown in similar places to *A. heterophyllus*, but less common, and not very successful in Uganda, see Dale, Introd. Trees Uganda: 10 (1953).

Broussonetia papyrifera (L.) Vent.; Paper Mulberry: native of tropical and subtropical Asia. Introduced to Uganda about the middle of the century for paper production, see Dale, Introd. Trees Uganda: 14 (1953), also to Tanzania at Amani and Lushoto.

Castilla elastica Sessé; Panama Rubber; native of Central and western South America. Tried at Amani (see T.T.C.L.: 351 (1949)), in Zanzibar (see U.O.P.Z.: 182 (1949)) and at Entebbe (see Dale, Introd. Trees Uganda: 20 (1953)), but not developed commercially; both subsp. *elastica* and subsp. *costaricana* (Liebm.) C.C. Berg, with more appressed hairs on the underside of the leaves, have been introduced, see C.C. Berg in Fl. Neotropica 7: 94–100 (1972).

Plants herbaceous (or succulent) . . . . **10. Dorstenia**
Plants woody:
  Inflorescences urceolate; glandular, ± waxy, spots on the lamina beneath at the base of the midrib or in the axils of at least the main basal lateral veins . . . . . **11. Ficus**
  Inflorescences not urceolate; glandular spots on the lamina absent:
    Stipules fully amplexicaul (leaving annular scars):
      Stipules connate; inflorescences bisexual; staminate flowers without perianth . . . . **9. Trilepisium**
      Stipules free; inflorescences unisexual, or if bisexual, then the staminate flowers with a perianth:
        Inflorescences globose- to obovoid-capitate **5. Treculia**
        Inflorescences discoid- to turbinate-capitate or uniflorous:

*See also Cecropiaceae, p. 87

**With *Dorstenia* by M.E.E. Hijman (University of Utrecht). Both authors received grants from the Netherlands Organization for Advancement of Pure Research (ZWO) to carry out this study.

Inflorescences with only basally attached involucral bracts . . . . **7. Mesogyne**
Inflorescences with peltate interfloral bracts and peripheral basally attached bracts **8. Bosquieopsis**
Stipules semi-amplexicaul to lateral:
Plants armed with spines . . . . **3. Maclura**
Plants without spines:
Inflorescences bisexual . . . . **10. Dorstenia**
Inflorescences unisexual:
Staminate inflorescences discoid-capitate and involucrate; pistillate inflorescences uniflorous, involucrate, flower adnate to the receptacle . . . . **6. Antiaris**
Staminate inflorescences spicate; pistillate inflorescences globose-capitate or uniflorous with the flower free:
Lamina beneath densely puberulous on the vein-reticulum; both staminate and pistillate inflorescences spicate . . . . **2. Milicia**
Lamina beneath sparsely puberulous to glabrous; staminate inflorescence spicate and pistillate inflorescences ± globose or uniflorous:
Lamina subtriplinerved; peduncle of the staminate inflorescence 0.3–2 cm. long; pistillate inflorescences several-flowered **1. Morus**
Lamina pinnately veined; peduncle of staminate inflorescences up to 0.2 cm. long; pistillate inflorescences uniflorous **4. Sloetiopsis**

1. MORUS

L., Sp. Pl.: 986 (1753) & Gen. Pl., ed. 5: 424 (1754); C.C. Berg in B.J.B.B. 47: 335 (1977)

Trees, dioecious; shoot-apices shed. Leaves distichous, subtriplinerved; stipules lateral, free. Inflorescences bracteate. Staminate inflorescences spicate; tepals 4, basally connate; stamens 4, inflexed in bud; pistillode present. Pistillate inflorescences capitate; tepals 4, basally connate; ovary free; stigmas 2, filiform, subequal in length. Fruiting perianth enlarged, fleshy, greenish to yellow; fruit free, somewhat drupaceous. Seed small, with endosperm; cotyledons flat, equal, plane.

The genus comprises 10–15 species in temperate to subtropical regions of the Old and New Worlds; the only African species belongs to the tropical lowland flora.

M. australis Poir. (*M. indica* sensu auctt., *non* L.), the cultivated mulberry, has been grown quite widely in the midlands and highlands of Kenya and Tanzania, also in the eastern and northern parts of Uganda. *M. alba* L., the White Mulberry, has been introduced to Amani (*Greenway* 1757!), and differs by the stigmas sessile (not on a style) 1–1.5 mm. long and the lamina usually smooth (rather than ± scabrous) above.

M. mesozygia *Stapf* in Journ. de Bot., sér. 2, 2: 99 (1909); Rendle in F.T.A. 6(2): 21 (1916); Hauman in F.C.B. 1: 55 (1948); F.P.S. 2: 273 (1952); F.W.T.A., ed. 2, 1: 594 (1958); C.C. Berg in B.J.B.B. 47: 337, fig. 16 (1977) & in Fl. Cameroun 28: 6, t. 1 (1985). Type: Ivory Coast, Zaranou, *Chevalier* 16267 (P, lecto.!, K, isolecto.!)

Tree up to 35 m. tall. Lamina chartaceous to subcoriaceous, elliptic to oblong, ovate or obovate, 3–13 × 2–8 cm., apex acuminate to subacute, base cordate to obtuse, margin crenate to serrate; upper surface pubescent on the main veins, lower surface pubescent in the axils of lateral veins; lateral veins 4–7 pairs, the basal pair strong, the others much weaker, departing from the upper part of the midrib, tertiary venation partly scalariform; petiole 0.5–2 cm. long; stipules ± 0.5 cm. long, caducous. Staminate inflorescences: spike 1–2.5 cm. long, ± 0.8 cm. diameter; peduncle 0.3–2 cm. long. Pistillate inflorescences ± 0.5(–1 in fruit) cm. in diameter; peduncle 0.4–2 cm. long; stigmas 3–5 mm. long. Fruit ellipsoid to subglobose, ± compressed, 5 × 3–5 mm. Fig. 1.

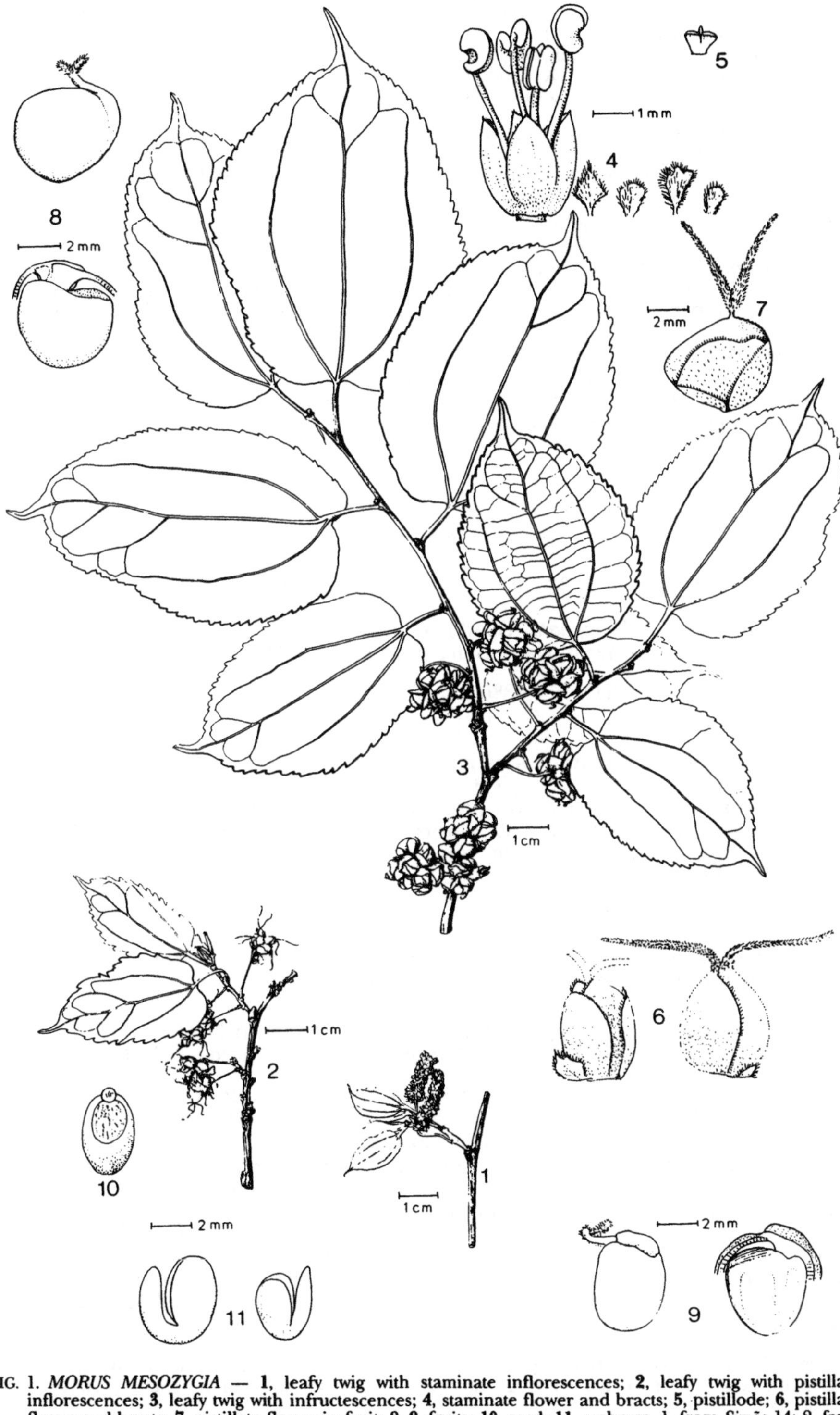

FIG. 1. *MORUS MESOZYGIA* — **1**, leafy twig with staminate inflorescences; **2**, leafy twig with pistillate inflorescences; **3**, leafy twig with infructescences; **4**, staminate flower and bracts; **5**, pistillode; **6**, pistillate flower and bracts; **7**, pistillate flower in fruit; **8**, **9**, fruits; **10**, seed; **11**, embryos. 1, from *Simão* 14; 2, from *Fanshawe* 9319; 3, 8–11, from *Simão* 233; 4,5, from *Andrada* 1447; 6, from *Gomes e Sousa* 1862; 7, from *Espirito Santo* 1961. Drawn by E.H. Hupkens van der Elst and W. Scheepmaker.

UGANDA. Bunyoro District: Budongo Forest, Feb. 1932, *C.M. Harris* 48! & *Dawe* 785 & Siba Forest Reserve, 28 Sept. 1962, *Styles* 101!
KENYA. N. Kavirondo District: Kakamega, *Wormald* 25!
TANZANIA. Mwanza District: Geita, 7 Apr. 1953, *Marshall* 2/53!; Tanga District: Kihuhwi, 14 Nov. 1944, *Greenway* 7037!; Morogoro District: E. Uluguru Mts., Uponda, 20 Feb. 1948, *Gilchrist* in *F.H.* 2285!
DISTR. **U** 2, 4; **K** 5 ; **T** 1, 3, 4, 6, 8; Senegal to SW. Ethiopia, south to Angola, Zambia, Mozambique and South Africa (Natal)
HAB. Rain-forest and drier evergreen forest, sometimes with *Milicia* and *Antiaris*, but apparently relatively uncommon; 450–1600 m.

SYN. *Celtis lactea* Sim, For. Fl. Port. E. Afr.: 97, t. 91 (1909); Rendle in F.T.A. 6(2): 4 (1916). Type: Mozambique, Ouisico, *Sim* 5299 (K, holo.!)
Morus lactea (Sim) Mildbr. in N.B.G.B. 8: 243 (1922); Peter, F.D.O.-A. 2: 67 (1932); Hauman in F.C.B. 1: 55 (1948); T.T.C.L.: 362 (1949); I.T.U., ed. 2: 263, photo. 43 (1952); J. Leroy in Journ. Agric. Trop. Bot. Appliq. 2: 677, t. 6 (1955); K.T.S.: 324 (1961); Hamilton, Ug. For. Trees.: 94 (1981)
M. mesozygia Stapf var. *lactea* (Sim) A. Chev. in Rev. Bot. Appliq. 29: 72 (1949)

NOTE. The wood is strong, beautiful and easily worked.
Although Sim's Forest Flora of Portuguese East Africa is dated Jan. 1909 in the preface it was not received at Kew until July 1909 after publication of the first part of the Journal de Botanique, dated Apr. 1909.

2. MILICIA

Sim, For. Fl. Port. E. Afr.: 97 (1909); C.C. Berg in B.J.B.B. 52: 226 (1982)

Maclura Nuttall sect. *Chlorophora* (Gaud.) Baillon, Hist. Pl. 6: 193 (1875–76), pro parte; Corner in Gard. Bull., Singapore 19: 236 (1962)

[*Chlorophora* sensu G.P. 3(1): 363 (1880) et auct. Afr. mult. *non* Gaud.; C.C. Berg in B.J.B.B. 47: 347 (1977)]

Trees, dioecious. Leaves distichous, pinnately veined; stipules semi-amplexicaul, free. Inflorescences in the leaf-axils, spicate, bracteate. Staminate flowers: tepals 4, basally connate; stamens 4, inflexed in bud; pistillode present. Pistillate flowers 5–15; tepals 4, basally connate; ovary free; stigmas 2, filiform, very unequal in length. Fruiting perianth enlarged, ± fleshy, greenish; fruit free, somewhat drupaceous. Seed small, with endosperm; cotyledons thin, equal, plane.

Two species in tropical Africa.

The genus *Chlorophora* was based by Gaudichaud on *Morus tinctoria* L., an American species now referred to the genus *Maclura* (see p. 6).

M. excelsa (*Welw.*) *C.C. Berg* in B.J.B.B. 52: 227 (1982) & in Fl. Cameroun 28: 9, t. 2 (1985). Type: Angola, Golungo Alto, *Welwitsch* 1559 ♀ (LISU, holo.!, B, BM, G, K, P, iso.!)

Tree up to 30(–50) m. tall. Lamina coriaceous, when juvenile chartaceous, elliptic to oblong, 6–20(–33) × 3.5–10(–12) cm., apex acuminate, base cordate to obtuse, margin subentire to repand, when juvenile serrate- to crenate-dentate, upper surface glabrous or puberulous to pubescent on the main nerves, when juvenile often scabridulous, lower surface densely puberulous on the vein-reticulum, pubescent, puberulous or almost glabrous on the main veins, when juvenile the whole surface hirtellous to tomentose; lateral veins 10–22 pairs, tertiary venation partly scalariform; petiole 1–5 cm. long; stipules 0.5–5 cm. long, caducous. Staminate inflorescences: spike 8–20 cm. long, ± 0.5 cm. in diameter; peduncle 0.5–2.5 cm. long. Pistillate inflorescences: spike 2–3(–5 in fruit) cm. long, 0.5(–1.5 in fruit) cm. in diameter; stigmas up to 7 mm. long. Fruit ellipsoid, 2.5–3 mm. long. Fig. 2.

UGANDA. Bunyoro District: Budongo Forest, Waibira, 20 Sept. 1962, *Styles* 55!; Mengo District: Entebbe, Nov. 1905, *E. Brown* 354! & E. Mabira Forest Reserve, Nagojje Forest Station, 20 Nov. 1962, *Styles* 223!
KENYA. C. Kavirondo District: Nangina–Sio road, 26 July 1945, *R.L. Davidson* 453; Kwale District: Shimba Hills, Mar. 1937, *Dale* in *F.D.* 1072!; Lamu District: Witu, *C.W. Elliot* 212!

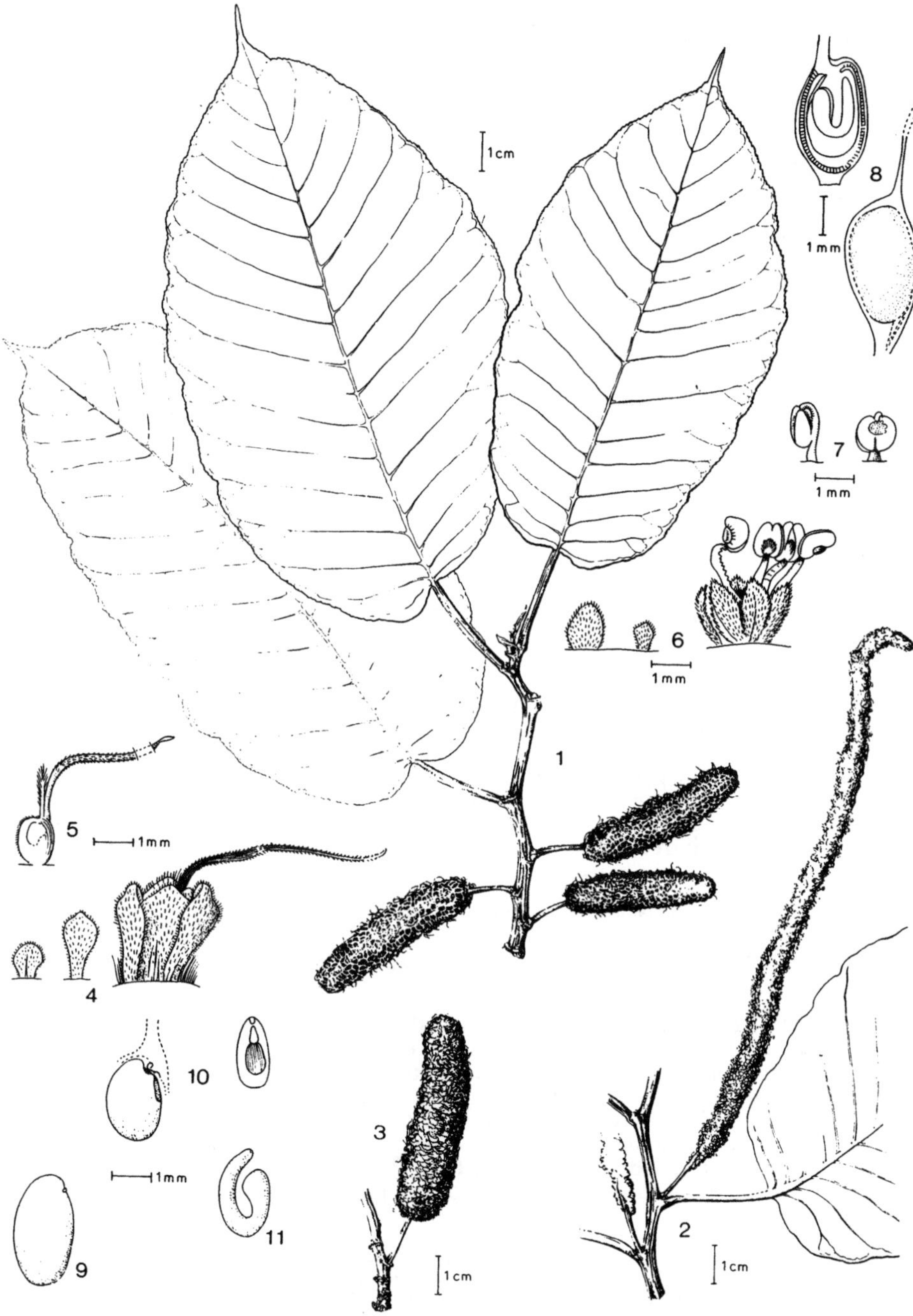

FIG. 2. *MILICIA EXCELSA* — **1**, leafy twig with pistillate inflorescences; **2**, staminate inflorescence; **3**, infructescence; **4**, pistillate flower and bracts; **5**, pistil; **6**, staminate flower and bracts; **7**, stamen; **8**, fruit; **9**, endocarp-body; **10**, seed; **11**, embryo. 1, from *Simão* 152; 2, from *Barbosa* 2611; 3, from *Torre & Paiva* 9372; 4, 5, from *Léonard* 1049; 6, 7, from *Barbosa* 2278; 8–11, from *Simão* 654. Drawn by E.A. Hupkens van der Elst and W. Scheepmaker.

TANZANIA. Same District: Gonja, 20 Aug. 1954, *Hughes* 244!; Kilosa District: Kidodi, Oct. 1952, *Semsei* 960!; Rungwe District: Itamba, 20 Oct. 1932, *R.M. Davies* 646!; Zanzibar I., Dunga, 20 Dec. 1900, *Lyne* 6!

DISTR. **U** 1, 2, 4; **K** 4, 5, 7; **T** 1–4, 6–8; **Z**; **P**; Guinea Bissau to SW. Ethiopia, south to Angola, Mozambique, Malawi and E. Zimbabwe

HAB. A secondary tree of rain-forest, lowland evergreen forest, riverine and ground-water forest, sometimes left in old cultivated areas; 0–1350 m.

SYN. *Morus excelsa* Welw. in Trans. Linn. Soc., Bot. 27: 69, t. 23 (1869)

Maclura excelsa (Welw.) Bureau in DC., Prodr. 17: 231 (1873); Corner in Gard. Bull., Singapore 19: 257 (1962)

Chlorophora excelsa (Welw.) Benth. & Hook.f., G.P. 3(1): 363 (1880); Engl., E.M. 1: 3 (1898); Rendle in F.T.A. 6(2): 22 (1916); Peter, F.D.O.-A. 2: 69 (1932); Eggeling & Harris, Fifteen Ug. Timbers: 83, fig. 13, photo. 13 (1939); Hauman in F.C.B. 1: 56 (1948); T.T.C.L.: 351 (1949); I.T.U., ed. 2: 234, fig. 52, photo. 41 (1952); F.P.S. 2: 257, fig. 90 (1952); F.W.T.A., ed. 2, 1: 595 (1958); K.T.S.: 309, fig. 60 (1961); C.C. Berg in B.J.B.B. 47: 349, fig. 19 (1977); Hamilton, Ug. For. Trees: 94 (1981)

Milicia africana Sim, For. Fl. Port. E. Afr.: 97, t. 122 (1909). Type: Mozambique, *Sim* 5386 (not yet traced)

NOTE. Yields a fine heavy durable timber, known as iroko or mvule. Alan Hamilton (1981) considers this as the most valuable timber tree in East Africa.

3. MACLURA

Nuttall, Gen. N. Amer. Pl. 2: 233 (1818); C.C. Berg in Konink. Nederl. Akad. Weten., Ser. C, 89: 241–247 (1986), *nom. conserv.*

Chlorophora Gaud. in Freyc., Voy. Monde, Bot.: 508 (1830); G.P. 3(1): 363 (1880), excl. sp. Afr.

Cardiogyne Bureau in DC., Prodr. 17: 232 (1873); C.C. Berg in B.J.B.B. 47: 359 (1977)

Maclura Nuttall sect. *Cardiogyne* (Bureau) Corner in Gard. Bull., Singapore 19: 237 (1962)

Shrubs, treelets or climbers, dioecious, armed with spines. Leaves in spirals or sometimes subdistichous, pinnately veined; stipules lateral, free or connate spine-forming branchlets. Inflorescences in the axils of the leaves, globose-capitate, bracteate, with embedded yellow dye-containing glands in tepals and bracts. Tepals 4, partly connate. Stamens 4, inflexed in bud; pistillode present in staminate flower. Ovary free; stigmas 1 or 2 (but then unequal in length), filiform. Fruiting perianth enlarged, fleshy, yellow to orange; fruit free, somewhat drupaceous. Seed rather small; cotyledons thin, equal, plicate.

11 species in the Old and New Worlds; only one species in Africa.

M. africana (*Bureau*) *Corner* in Gard. Bull., Singapore 19: 257 (1962). Type: Tanzania, Zanzibar, *Boivin* (P, holo.!, B, iso.!)

Shrub or treelet up to 7 m. tall or a climber, with long weak branches and up to 10 cm. long branchlets ending in a spine. Lamina subcoriaceous, elliptic to lanceolate, rarely obovate or subcircular, 1.5–9 × 1–4.5 cm., apex obtuse to subacute to shortly acuminate or emarginate, base acute to obtuse, margin entire; upper surface glabrous or almost so, lower surface sparsely puberulous; lateral veins 4–12 pairs, tertiary venation reticulate; petiole 0.3–3 cm. long; stipules up to 0.2 mm. long, persistent. Staminate inflorescences ± 0.5–1.5 cm. in diameter; peduncle 0.5–2 cm. long. Pistillate inflorescence 0.5–0.8(–1.8 in rruit) cm. in diameter; peduncle 0.1–0.5 cm. long; stigmas up to 13 mm. long. Fruit ovoid, 6–7 mm. long. Fig. 3.

KENYA. Kwale District: Ukunda, 11 May 1953, *Bally* 8908! & Shimoni, 20 Aug. 1953, *Drummond & Hemsley* 3922!; Mombasa, 26 Jan. 1926, *R.M. Graham* in *F.D.* 1953!

TANZANIA. Tanga District: Kigombe Estate, Pangani road, 18 Oct. 1957, *Faulkner* 2073!; Morogoro District: Uluguru Mts., above Morningside, June 1953, *Semsei* 1241!; Rufiji District: Mafia I., Miskitini-Bweni, 12 Aug. 1937, *Greenway* 5056!; Zanzibar I., Marahubi, 9 July 1963, *Faulkner* 3218!

DISTR. **K** 6, 7; **T** 2, 3, 5, 6, 8; **Z**; south through Mozambique to NE. South Africa, and inland mainly through river basins to S. Malawi, SE. Zambia and E. Zimbabwe

HAB. Common along coast in bushland and dry evergreen forest, also extending inland mainly along rivers, sometimes in saline places; 0–720 m.

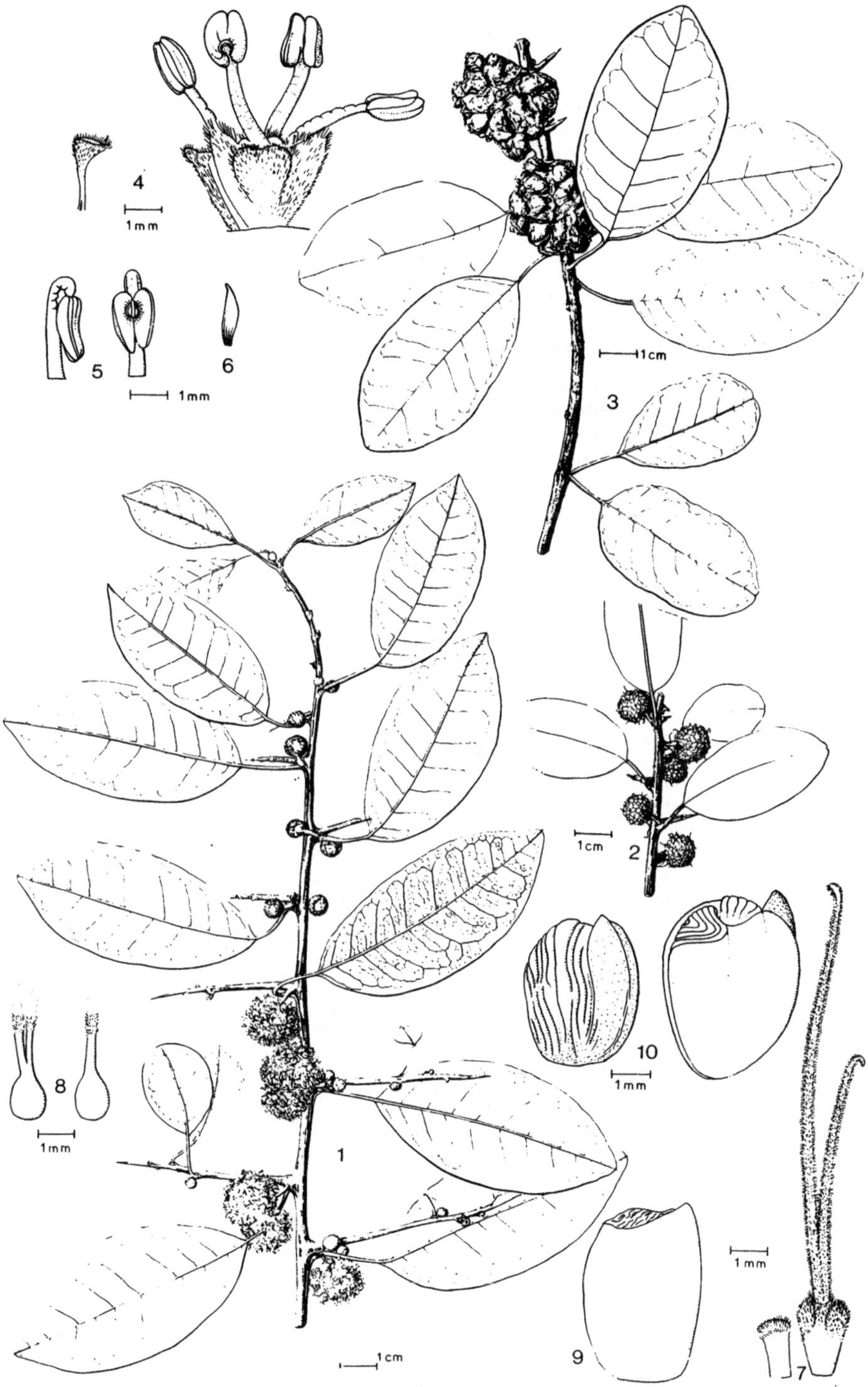

FIG. 3. *MACLURA AFRICANA* — **1**, leafy twig with staminate inflorescences; **2**, leafy twig with pistillate inflorescences; **3**, leafy twig with infructescences; **4**, staminate flower and bracts; **5**, stamens; **6**, pistillode; **7**, pistillate flower and bract; **8**, pistil; **9**, seed; **10**, embryo. 1, from *Torre* 7142; 2, from *Andrada* 1644; 3, 9, 10, from *Simão* 603; 4–6, from *Faulkner* 1610; 7, 8, from *Drummond & Hemsley* 2391. Drawn by E.H. Hupkens van der Elst and W. Scheepmaker.

SYN. *Cardiogyne africana* Bureau in DC., Prodr. 17: 233 (1873); Kirk in J.L.S. 9: 229 (1866) as '*Cudranea*'; Oliv. in Hook., Ic. Pl. 25, t. 2473 (1896); Engl., E.M. 1: 4 (1898); Rendle in F.T.A. 6(2): 24 (1916); Peter, F.D.O.-A. 2: 69 (1932); T.T.C.L.: 351 (1949); K.T.S.: 309 (1961); C.C. Berg in B.J.B.B. 47: 360, fig. 22 (1977)

NOTE. The immersed yellow glands occur below the thickened part of the tepals and bracts. Normally two glands occur, each beside the midvein, but may be irregularly present, reduced or, in some specimens, lacking. In pistillate flowers they may develop only after the fruit begins to mature.

4. SLOETIOPSIS

Engl. in E.J. 39: 573 (1907); C.C. Berg in B.J.B.B. 47: 363 (1977)

Neosloetiopsis Engl. in E.J. 51: 426 (1914)

Shrubs or treelets, dioecious or sometimes monoecious. Leaves distichous, pinnately veined; stipules semi-amplexicaul, free. Inflorescences in the leaf-axils, bracteate. Staminate inflorescences spicate; tepals 4, basally connate; stamens 4, inflexed in bud; pistillode present. Pistillate inflorescences uniflorous; tepals 4, free; ovary free; stigmas 2, filiform, subequal in length. Fruiting perianth enlarged, hardly fleshy, green. Fruit a free dehiscent drupe, the white fleshy exocarp pushing out the black endocarp-body (pyrene). Seed large, without endosperm; cotyledons thick, equal.

One species in tropical Africa, but its systematic position is uncertain. It fits in the genus *Streblus* Lour. sensu Corner, Gard. Bull., Singapore 19: 215 (1962), but that appears to be heterogenous and needs revision.

S. usambarensis *Engl.* in E.J. 39: 573, t. (1907) & V.E. 3(1): 20, t. 10 (1915); Rendle in F.T.A. 6(2): 77 (1916); Peter, F.D.O.-A. 2: 69 (1932); T.T.C.L.: 363 (1949); C.C. Berg in B.J.B.B. 47: 364, fig. 23 (1977) & in Fl. Cameroun 28: 12, t. 3 (1985). Type: Tanzania, Lushoto District, Mombo, *Engler* 3263 (B, holo.!)

Shrub or treelet up to 5 m. tall. Lamina subcoriaceous, oblong to elliptic or sometimes lanceolate, (1–)2–16 × (0.5–)1.5–6 cm., apex acuminate to subcaudate, base rounded to acute, margin crenate or serrate-dentate at least towards the apex or subentire; both surfaces glabrous or almost so; lateral veins 4–13 pairs, tertiary venation reticulate; petiole 0.2–0.7 cm. long; stipules 0.2–0.8 cm. long, often subpersistent. Staminate inflorescences: spike 0.5–5 cm. long, ± 0.4 cm. thick, subsessile (peduncle up to 1.5 mm.). Pistillate inflorescences: peduncle 0.2–0.3(–0.5 in fruit) cm. long; tepals ± 2(–5 in fruit) mm. long; stigmas (2–)6–8 mm. long. Fruit ± 10 mm. long; endocarp-body subglobose, ± 10 mm. in diameter, black. Fig. 4.

KENYA. Kwale District: Muhaka Forest, *Brenan, Gillett et al.* in *E.A.H.* 14534!; Kilifi District: Kaloleni–Kilifi, Cha Shimba [Chasimba], 30 Dec. 1970, *Faden, Evans & Msafiri* 70/950! & Tsagwa, Jibana, 16 Feb. 1979, *Gilbert & May* 5341!

TANZANIA. Pangani District: Pangani R. between Hale and Makinyumbe, 1 July 1953, *Drummond & Hemsley* 3125!; Morogoro District: Turiani, 22 Nov. 1955, *Milne-Redhead & Taylor* 7410! & Kimboza Forest Reserve, 27 Sept. 1971, *Pócs* 6466/G!

DISTR. **K** 7; **T** 3, 6; Guinée to Zaire, then coast of East Africa south to Mozambique, disjunct in several parts of its range

HAB. Forest, often riverine, in rocky places or on limestone outcrops; 30–700 m.

SYN. *Neosloetiopsis kamerunensis* Engl. in E.J. 51: 426, fig. 1 (1914); Rendle in F.T.A. 6(2): 78 (1916); Hauman in F.C.B. 1: 82 (1948); F.W.T.A., ed. 2, 1: 595 (1958). Type: Cameroun, N. of Moloundou, *Mildbraed* 4331 (B, holo.!)

5. TRECULIA

Decne. in Ann. Sci. Nat., sér. 3, 8: 108 (1847); C.C. Berg in B.J.B.B. 47: 378 (1977)

Trees, dioecious or sometimes monoecious. Leaves distichous or almost so, pinnately veined; stipules fully amplexicaul, free. Inflorescences unisexual, sometimes bisexual, in the leaf-axils and/or (especially the pistillate ones) on the older wood down to the trunk,

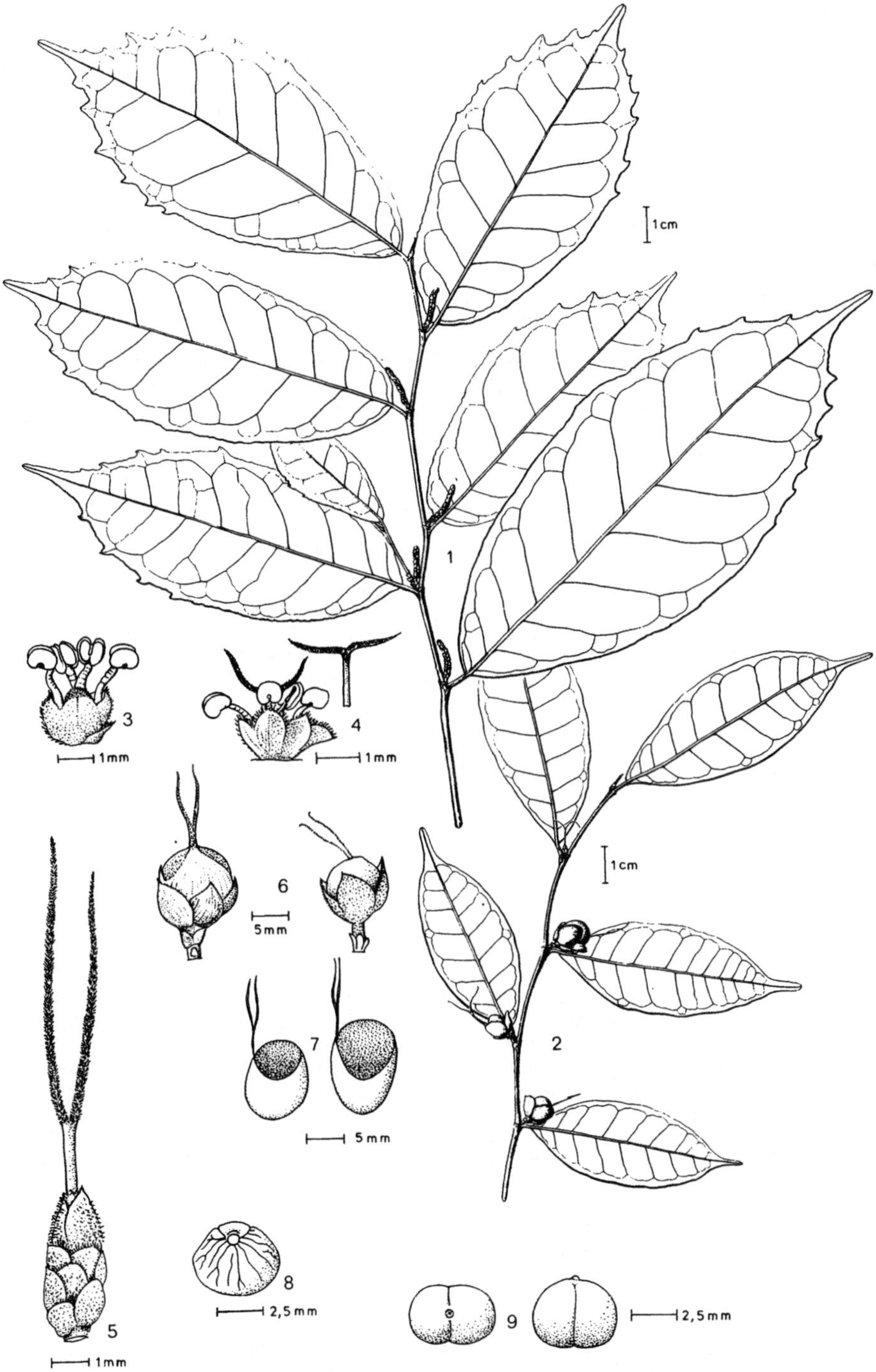

FIG. 4. *SLOETIOPSIS USAMBARENSIS* — **1**, leafy twig with young staminate inflorescences; **2**, leafy twig with young infructescences; **3**, staminate flower; **4**, staminate flower with bract and pistillode; **5**, pistillate inflorescence; **6**, infructescence; **7**, fruit; **8**, seed; **9**, embryo. 1, from *Simão* 336; 2, from *Gutzwiller* 2514; 3, from *de Wilde* 3542; 4, from *Hill* 296; 5, from *Breteler* 1865; 6–9, from *Breteler* 1507. Drawn by E.H. Hupkens van der Elst and W. Scheepmaker.

globose to obovoid-capitate, with a thick rachis and numerous peltate long-stipitate bracts. Staminate flowers: perianth 2–4(–5)-lobed; stamens 2–4, straight in bud; pistillode usually absent. Pistillate flowers without a perianth; stigmas 2, filiform, equal. Fruits somewhat drupaceous, embedded in the soft middle layer of the infructescence. Seed large, with remnants of endosperm; cotyledons unequal, curved, one thick, the other thin.

Three species in Africa and Madagascar.

T. africana *Decne.* in Ann. Sci. Nat., sér. 3, 8: 108, t. 3 (1847); Hook.f., Bot. Mag. 98, t. 5986 (1872); Engl., E.M. 1: 32, t. 12, 13 (1898); Hutch. in F.T.A. 6(2): 227 (1917); Hauman in F.C.B. 1: 90 (1948); T.T.C.L.: 363 (1949); I.T.U., ed. 2: 265 (1952); F.P.S. 2: 273, fig. 97 (1952); F.W.T.A., ed. 2, 1: 613 (1958); C.C. Berg in B.J.B.B. 47: 382, fig. 28, 30 (1977) & in Fl. Cameroun 28: 16, t. 4 (1985). Type: 'Senegambia', *Heudelot* (P, holo.!)

Tree up to 30(–50) m. tall. Lamina coriaceous, oblong or lanceolate to subovate, sometimes elliptic or ovate, (5–)10–25(–50) × (2.5–)4–12(–20) cm., apex acuminate or sometimes subacute, base obtuse to subcordate, sometimes subacute, margin entire to faintly repand; upper surface glabrous or almost so, lower surface sparsely puberulous on the main veins; lateral veins (8–)10–18 pairs, tertiary venation mainly reticulate; petiole 0.2–1.5 cm. long; stipules 1–1.8 cm. long, glabrous, puberulous or hirtellous, caducous. Inflorescences globose, ellipsoid or obovoid, 2.5–10 cm. in diameter; peduncle up to 0.4 cm. long. Stigmas 3.5(–10) mm. long. Infructescences subglobose, up to 30 cm. in diameter; fruit ellipsoid to oblong, 10–15 mm. long. Figs. 5 & 6, p. 12.

UGANDA. Bunyoro District: Budongo Forest, Pabidi [Pabeddi], June 1932, *C.M. Harris* 133!; Mengo District: Busiro, *Dawe* 145! & Namanve Forest, Apr. 1932, *Eggeling* 587!
TANZANIA. Bukoba District: Munene Forest, July 1951, *Eggeling* 6254!; Kigoma District: Mugombazi, 1 Sept. 1959, *Harley* 9503!; Kilosa District: Kidodi, Msowelo R., Oct. 1952, *Semsei* 948!
DISTR. **U** 2, 4; **T** 1, 3, 4, 6–8; tropical Africa and Madagascar
HAB. Forest, generally riverine; 0–1200 m.

NOTE. The East African material is all referable to subsp. *africana* var. *africana*. Var. *mollis* (Engl.) J. Léon. (fig. 5/1), with a dense tomentum on the leaves, twigs and stipules, is only known from Nigeria, Cameroun, Gabon and Zaire. Subsp. *madagascarica* (N.E. Br.) C.C. Berg has several varieties in Madagascar and is not easily separated as a whole from the continental race.

6. ANTIARIS

Leschen. in Ann. Mus. Hist. Nat. Paris 16: 478 (1810); C.C. Berg in B.J.B.B. 47: 308 (1977), *nom. conserv.*

Trees, monoecious or dioecious, with self-pruning lateral branches. Leaves distichous on the lateral branches, pinnately veined; stipules semi-amplexicaul, free. Inflorescences on minute spurs unisexual, in the leaf-axils or just below the leaves, involucrate. Staminate inflorescences pedunculate; flowers numerous; tepals 2–7, free; stamens 2–4, straight in bud; pistillode absent. Pistillate inflorescences sessile or pedunculate, uniflorous; perianth partly adnate to the receptacle, 4-lobed; ovary adnate to the perianth; stigmas 2, band-shaped, equal. Fruit forming a drupaceous whole with the enlarged fleshy orange to scarlet receptacle. Seed large, without endosperm; cotyledons thick, equal.

One species in the Old World tropics. The three varieties are not very clear-cut morphologically. Juvenile specimens are more similar and can often hardly be told apart.

A. toxicaria *Leschen.* in Ann. Mus. Hist. Nat. Paris 16: 478, t. 22 (1810); I.T.U., ed. 2: 233 (1952); K.T.S.: 308 (1961); Corner in Gard. Bull. Singapore 19: 244 (1962); C.C. Berg in B.J.B.B. 47: 309, 310 (1977); Hamilton, Ug. For. Trees: 93 (1981); C.C. Berg in Fl. Cameroun 28: 106, t. 36, 37 (1985). Type: Java; no specimen cited

Tree up to 40(–60) m. Lamina coriaceous, when juvenile often chartaceous, elliptic to oblong or ± obovate, when juvenile often lanceolate, (2–)6–15(–32) × (1.5–)3–12 cm., apex shortly acuminate to obtuse or subacute, base obtuse to subcordate or, less often, subacute, margin subentire or denticulate, when juvenile often dentate, puberulous to

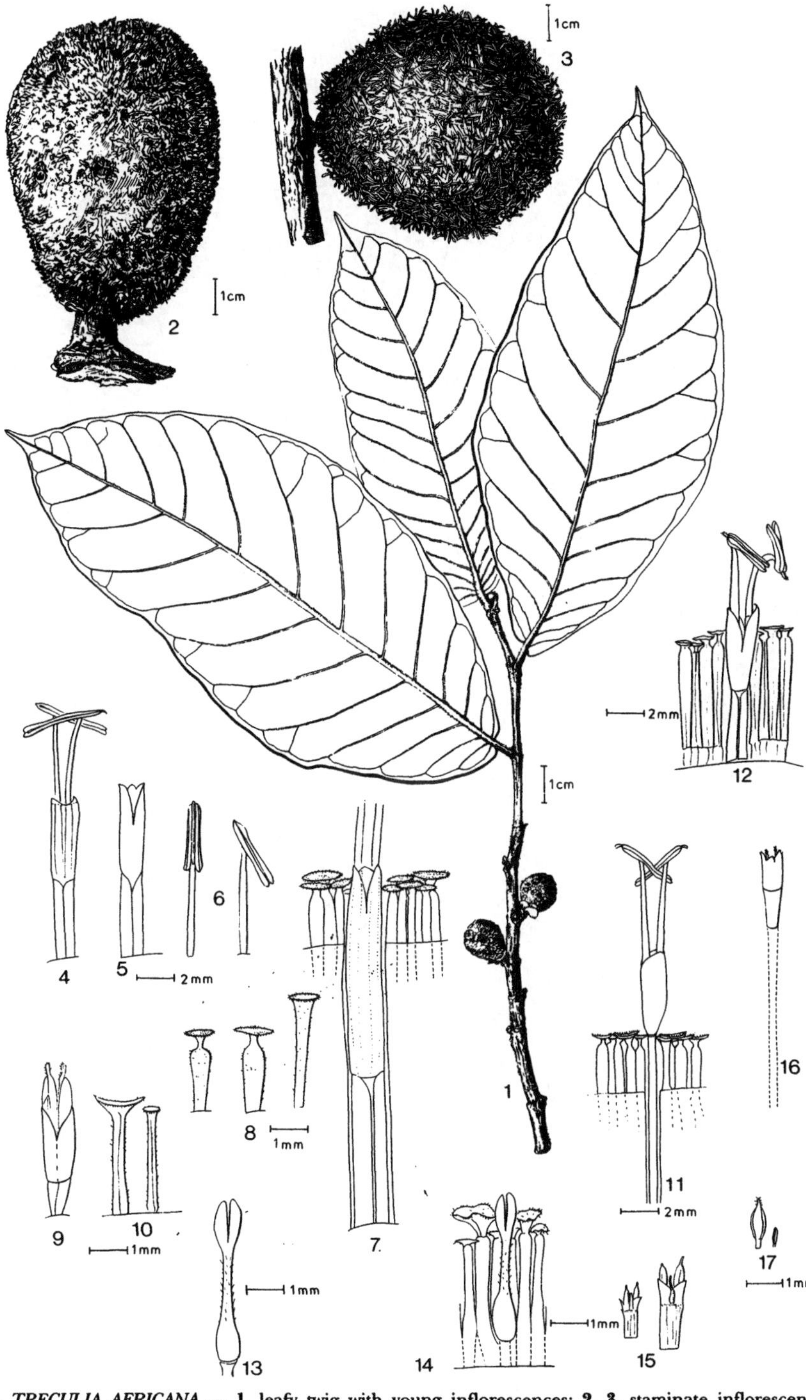

FIG. 5. *TRECULIA AFRICANA* — **1**, leafy twig with young inflorescences; **2**, **3**, staminate inflorescence; **4**, staminate flower; **5**, perianth; **6**, stamens; **7**, staminate flower and bracts; **8**, bracts; **9**, staminate flower from another plant; **10**, bracts; **11**, **12**, staminate flower and bracts; **13**, young pistillate flower; **14**, young pistillate flower and bracts; **15**, abortive staminate flower; **16**, abortive staminate flower in pistillate inflorescence; **17**, stamen and pistillode of abortive staminate flower. 1, from *Mpom* 134; 2, 11, from *Le Testu* 3848; 3, from *de Wilde* 2605; 4–8, from *de Wilde* 2662; 9, 10, from *Callens* 2886; 12, from *Zenker* 2525; 13–16, from *Le Testu* 3831; 16, 17, from *Leeuwenberg* 10217. Drawn by E.H. Hupkens van der Elst and W. Scheepmaker.

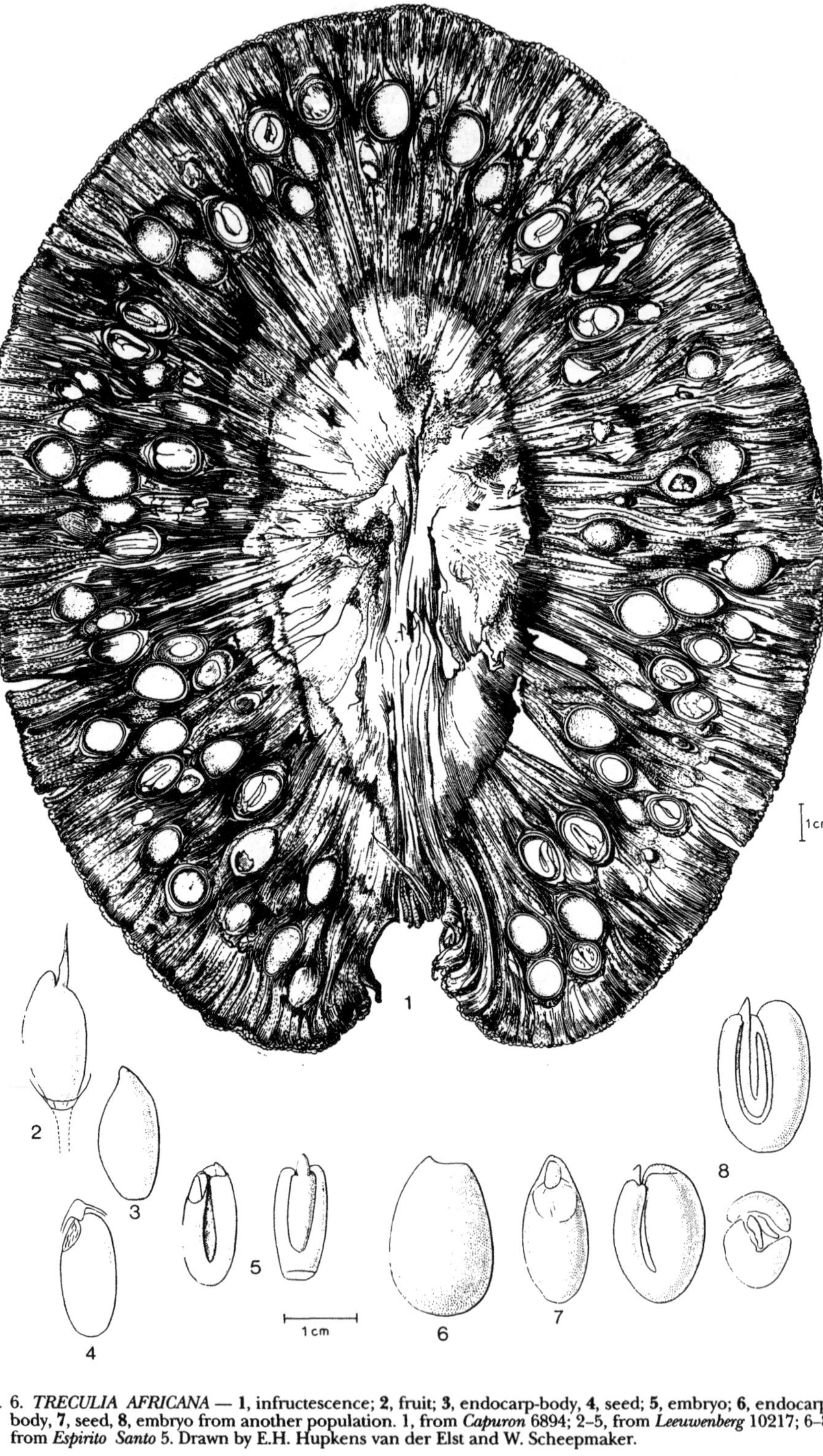

FIG. 6. *TRECULIA AFRICANA* — **1**, infructescence; **2**, fruit; **3**, endocarp-body, **4**, seed; **5**, embryo; **6**, endocarp-body, **7**, seed, **8**, embryo from another population. 1, from *Capuron* 6894; 2–5, from *Leeuwenberg* 10217; 6–8, from *Espirito Santo* 5. Drawn by E.H. Hupkens van der Elst and W. Scheepmaker.

densely ± stiffly hairy; lateral veins (5–)7–14 pairs, tertiary venation partly scalariform; petiole 3–10 mm. long; stipules 3–10(–15) mm. long, caducous. Staminate inflorescences 0.6–1.2(–2) cm. in diameter; peduncle 0.5–1.5(–1.8) cm. long. Pistillate inflorescences 3–4 cm. in diameter, sessile or with a peduncle up to 3–6 mm. long; stigmas (2–)5–8(–10) mm. long. Infructescences ellipsoid, sometimes ovoid or globose, 1–1.5 × 0.8–1 cm.

subsp. **welwitschii** (*Engl.*) *C.C. Berg* in B.J.B.B. 48: 466 (1978); Troupin, Fl. Pl. Lign. Rwanda: 434 (1982); C.C. Berg in Fl. Cameroun 28: 106, t. 36 (1985). Type: Angola, Golungo Alto, *Welwitsch* 2593 (B, lecto.!, BM, G, K, LISU, P, isolecto.!)

a. var. **welwitschii** (*Engl.*) *Corner* in Gard. Bull., Singapore 19: 248 (1962)

Lamina subcoriaceous, margin subentire; upper surface smooth, on the midrib puberulous to hirtellous, lower surface smooth or on the midrib sparsely appressed puberulous, occasionally hirtellous; only the midrib and lateral veins prominent beneath, the smaller veins plane or almost so.

UGANDA. Mengo District: Kampala, Dec. 1921, *Snowden* 728! & Lwamunda [Lyamunda] Forest Reserve, near Butu, 14 Nov. 1962, *Styles* 208! & Entebbe golf-links, Feb. 1932, *Eggeling* 414!
TANZANIA. Bukoba District, without precise locality or date, *Gillman* 621!
DISTR. **U** 4; **T** 1; westwards to Sierra Leone and south to Angola and Zambia
HAB. Rain-forest; 1100–1200 m.

SYN. *A. welwitschii* Engl. in E.J. 33: 118 (1902); Hutch. in F.T.A. 6(2): 224 (1917); Hauman in F.C.B. 1: 93, photo. 6 (1948); F.W.T.A., ed. 2, 1: 613 (1958)
A. toxicaria Leschen. subsp. *africana* (Engl.) C.C. Berg var. *welwitschii* (Engl.) C.C. Berg in B.J.B.B. 47: 316, fig. 10/1–3 (1977)

b. var. **africana** *A. Chev.*, Veg. Ut. Afr. Trop. Fr. 5: 259 (1909); C.C. Berg in Fl. Cameroun 28: 110, t. 37 (1985). Type: Ivory Coast, Dabou, *Chevalier* 16217 (P, lecto.!, K, isolecto.!)

Lamina coriaceous or less often subcoriaceous, when dry ± brittle, margin subentire to denticulate; upper surface scabrous to scabridulous, hirtellous to hispidulous; lower surface ± scabrous, mostly hirtellous to hispidulous; beneath also the smaller veins prominent. Fig. 7.

UGANDA. Acholi District: Labworomor [Labworomo] Hill, Jan. 1939, *St. Clair Thompson* 1722!
DISTR. **U** 1; Senegal to NE. Zaire and S. Sudan
HAB. Wooded grassland; 1000–1100 m.

SYN. *A. africana* Engl. in E.J. 33: 119 (1902); Hutch. in F.T.A. 6(2): 223 (1917); F.P.S. 2: 257 (1952); F.W.T.A., ed. 2, 1: 612 (1958). Type: Togo, near Lome, *Warnecke* 336 (B, lecto.!, E, G, K, L, P, isolecto.!)
A. toxicaria Leschen. subsp. *africana* (Engl.) C.C. Berg var. *africana*; C.C. Berg in B.J.B.B. 47: 314, fig. 9 (1977)

NOTE. In drier habitats than var. *welwitschii*, but intermediates occur in areas of geographical proximity or overlap.

c. var. **usambarensis** (*Engl.*) *C.C. Berg* in B.J.B.B. 47: 318, fig. 10/4, 5 (1977) & in B.J.B.B. 48: 466 (1978). Type: Tanzania, Usambara Mts., near Derema, *Scheffler* 216 (B, holo.!, K, iso.!)

Lamina subcoriaceous, margin subentire, upper surface smooth to scabridulous, puberulous at least on the midrib, or glabrous, lower surface scabridulous to sometimes scabrous, hirtellous to hispidulous to puberulous on the veins; midrib, lateral veins and part of the smaller veins prominent beneath, the reticulum plane or almost so.

UGANDA. Bunyoro District: Budongo Forest, Feb. 1932, *C. M. Harris* 45!; Busoga District: Lolui I., 14 May 1964, *G. Jackson* U83!; Mbale District: Elgon, Feb. 1940, *St. Clair Thompson* in *Eggeling* 3957!
KENYA. N. Kavirondo District: Kakamega Forest, May 1933, *Dale* in *F.D.* 3128! & Kakamega Forest Station, 21 Dec. 1967, *Perdue & Kibuwa* 9405!; Kwale District: Mrima Hill, 16 Jan. 1964, *Verdcourt* 3937!
TANZANIA. Biharamulo District: Katoke Mission, 4 June 1956, *Gane* 74!; Kilosa District: Kidodi, Oct. 1952, *Semsei* 966!; Ulanga District: Kiberege, 22 July 1959, *Haerdi* 299/0!
DISTR. **U** 2, 3; **K** 5, 7; **T** 1–3, 6; E. Zaire
HAB. Rain-forest or at least wetter evergreen forests, riverine or semi-swamp forest; 10–1700 m.

SYN. *A. usambarensis* Engl. in E.J. 33: 119 (1902); Hauman in F.C.B. 1: 94 (1948); T.T.C.L.: 349 (1949); Troupin, Fl. Rwanda 1: 134 (1978)
[*A. challa* sensu Engl., V.E. 3(1): 33, t. 20C (1915), *non* (Schweinf.) Engl. sensu stricto]
A. toxicaria Leschen. subsp. *africana* (Engl.) C.C. Berg var. *usambarensis* (Engl.) C.C. Berg in B.J.B.B. 47: 38, fig. 10 (1977)

NOTE. Intermediate in morphological features between typical var. *welwitschii* and *africana*, but rather different from the intermediates found between those variants in W. Africa

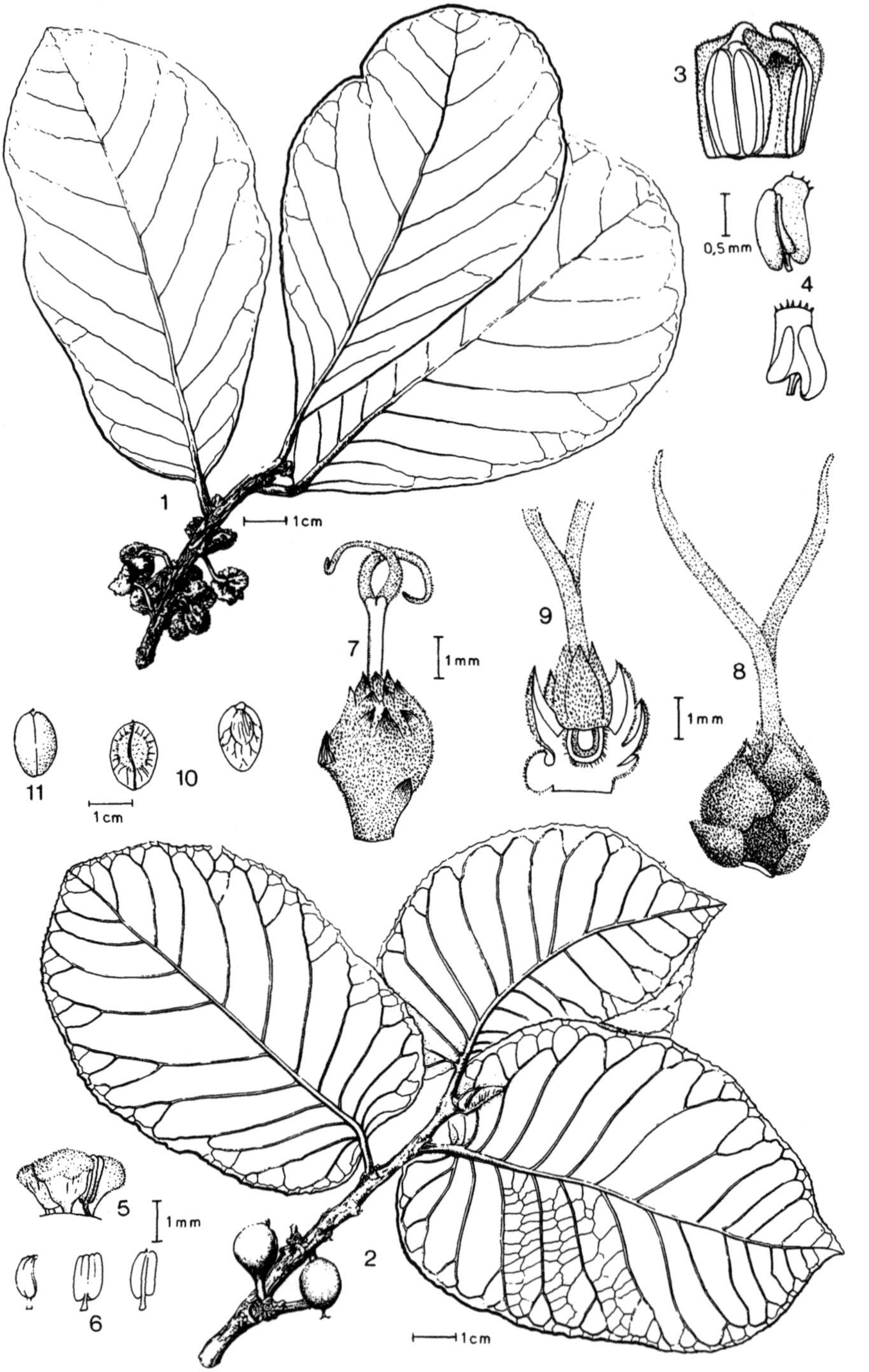

FIG. 7. *ANTIARIS TOXICARIA* subsp. *WELWITSCHII* var. *AFRICANA* — **1**, leafy twig with staminate inflorescences; **2**, leafy twig with pistillate inflorescences; **3**, staminate flower; **4**, stamens; **5**, staminate flower; **6**, stamens from another plant; **7**, **8**, **9**, pistillate flowers; **10**, seed; **11**, embryo. 1, 10, 11, from *Kersting* A. 571; 2, from *Espirito Santo* 1741; 3, 4, 7, from *Dalziel* 177; 5, 6, from *de Wilde* 1373; 8, 9, from *d'Orey* 296. Drawn by E.H. Hupkens van der Elst and W. Scheepmaker.

A. challa (Schweinf.) Engl., based originally on material from the Yemen, clearly belongs to *A. toxicaria*, but without fruits its infraspecific position is unclear.

DISTR. (of species as a whole). **U** 1–4; **K** 5, 7; **T** 1–3, 6; west to Senegal, north to S. Sudan, south to Angola and Zambia, with two subspecies in Madagascar and one in Asia

NOTE. Apart from local uses the wood has some commercial value for plywood and veneer.

7. MESOGYNE

Engl. in E.J. 20: 147 (1894); C.C. Berg in B.J.B.B. 47: 323 (1977)

Trees or shrubs, monoecious, probably with self-pruning lateral branches. Leaves distichous at least on the lateral branches, pinnately veined; stipules fully amplexicaul, free. Inflorescences in the leaf-axils, unisexual, discoid-capitate, involucrate. Staminate inflorescences pedunculate, with 3–4 involucral bracts; flowers several; sepals 2–4 basally connate; stamens 2–4, straight in bud; pistillode usually absent. Pistillate inflorescences sessile, uniflorous, with several involucral bracts; flowers partly adnate to the receptacle; perianth 3–4-lobed; ovary adnate to the perianth; stigmas 2, band-shaped, equal. Fruit forming a drupaceous whole with the enlarged fleshy red receptacle. Seed large, without endosperm; cotyledons very unequal.

One species in tropical Africa.

M. insignis *Engl.* in E.J. 20: 148, t. 5M–U (1894) & E.M. 1: 30, t. 11 (1898); Hutch. in F.T.A. 6(2): 222 (1917); Peter, F.D.O.-A. 2: 81 (1932); T.T.C.L.: 362 (1949); C.C. Berg in B.J.B.B. 47: 324, fig. 13 (1977). Type: Tanzania, Lushoto District, near Ngwelo, *Holst* 2290 (B, holo.!)

Shrubs or trees up to 15 m. tall. Lamina subcoriaceous to coriaceous, oblong to subobovate, sometimes elliptic or lanceolate, 5–26 × 1.5–9.5 cm., apex acuminate, base acute to obtuse, margin entire or faintly crenate-dentate; upper surface glabrous, lower surface sparsely puberulous, glabrescent; lateral veins 9–14 pairs, tertiary venation mostly reticulate; petiole 0.3–1(–1.5) cm. long; stipules 0.3–1.1 cm. long, caducous. Staminate inflorescences 2–4 mm. in diameter; peduncle 2–12 mm. long. Pistillate inflorescences 1–2 mm. in diameter; involucral bracts ± 12–16; stigmas 2.5–4 mm. long. Infructescences 2.5–3 × 1.5–2 cm. Fig. 8, p. 16.

TANZANIA. Lushoto District: Mt. Bomole, 21 Feb. 1950, *Verdcourt* 81! & Amani, Apr. 1951, *Parry* 6!; Morogoro District: Uluguru Mts., Tegetero, 20 Mar. 1953, *Drummond & Hemsley* 1692!

DISTR. **T** 3, 6; S. Tomé

HAB. Rain-forest; 500–1300 m.

NOTE. The material from S. Tomé differs in minor respects — longer (2.5–4 mm.) strap-shaped stigmas, less deeply divided perianth of the staminate flower, in the somewhat shorter (2–4 mm.) peduncle of the staminate inflorescence, and often in the somewhat more coriaceous leaves — but is too sparse to evaluate the significance of the variation.

8. BOSQUEIOPSIS

De Wild. & Th. Dur. in Bull. Herb. Boiss., sér 2, 1: 839 (1901); C.C. Berg in B.J.B.B. 47: 293 (1977)

Trees or shrubs, monoecious or ? sometimes androdioecious. Leaves distichous, pinnately veined to subtriplinerved; stipules fully amplexicaul, free. Inflorescences in the leaf-axils or just below the leaves, bisexual or staminate, discoid- to subglobose- to turbinate-capitate; bracts interfloral (peltate) and marginal (basally attached). Staminate flowers several–many; tepals 3–4, connate; stamens (1–)2, inflexed in bud; pistillode present. Pistillate flower 1, central, partly adnate to the receptacle; perianth 4-lobed; ovary adnate to the perianth; stigmas 2, band-shaped, equal. Fruit forming a drupaceous whole with the enlarged fleshy orange to yellow receptacle, crowned with the remnants of staminate flowers and bracts. Seed large, without endosperm; cotyledons thick, unequal.

One species in tropical Africa.

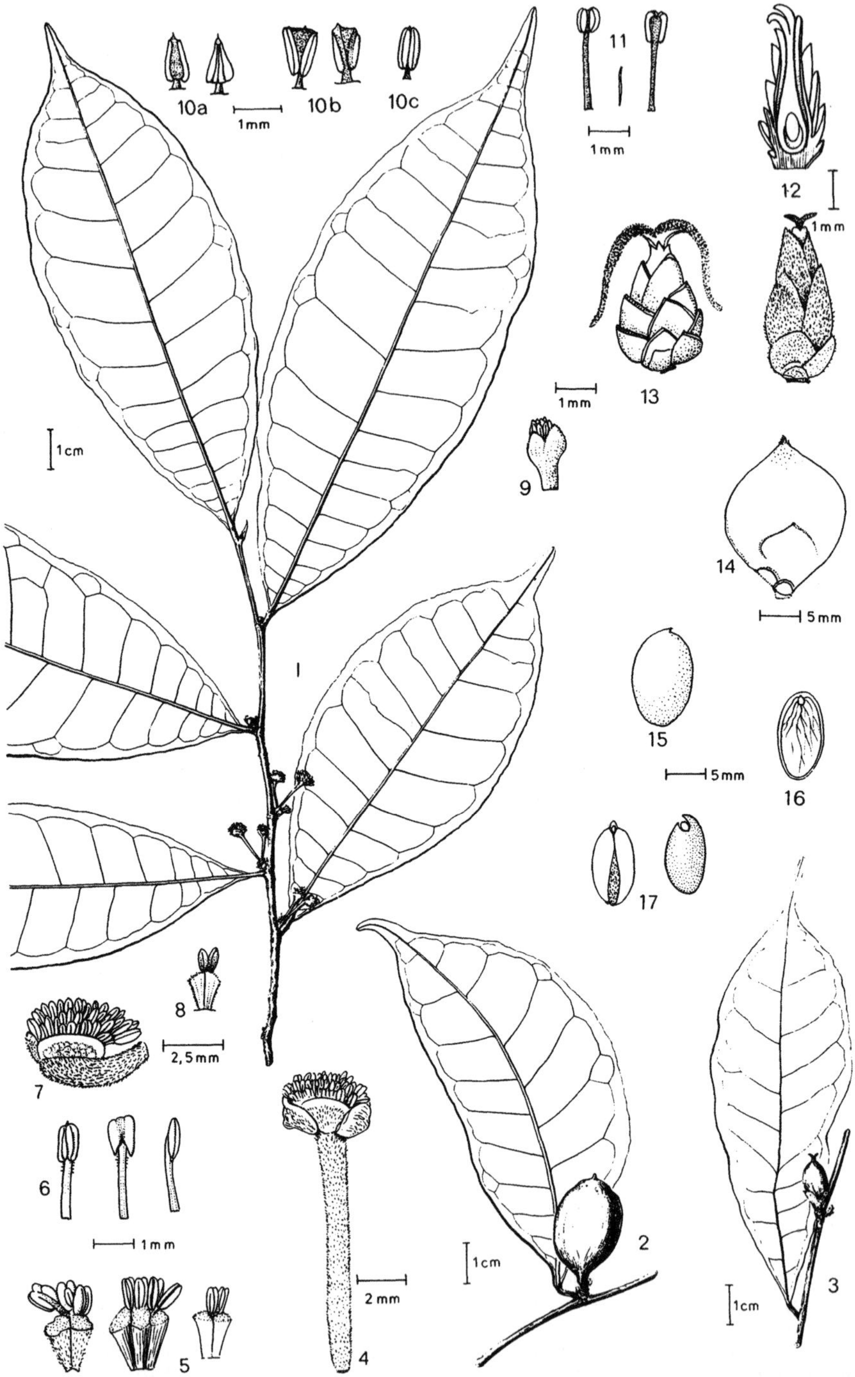

FIG. 8. *MESOGYNE INSIGNIS* — **1**, leafy twig with staminate inflorescences; **2**, leaf and infructescence; **3**, leaf and pistillate inflorescence; **4**, staminate inflorescence; **5**, staminate flowers; **6**, stamens; **7**, staminate inflorescence from another plant; **8, 9**, staminate flower; **10, 11**, stamens; **12, 13**, pistillate inflorescences; **14**, young infructescence; **15**, endocarp-body; **16**, seed. 1, from *Drummond & Hemsley* 1841; 2, from *Faulkner* 1346; 3, 12, from *Drummond & Hemsley* 1692; 4–6, from *Warnecke* 461; 7, 8, from *Drummond & Hemsley* 1694; 9–11, from *Quintas* 1062; 13, from *Chevalier* s.n.; 14–16, from *Semsei* 1447. Drawn by E.H. Hupkens van der Elst and W. Scheepmaker.

B. gilletii *De Wild. & Th. Dur.* in Bull. Herb. Boiss., sér. 2, 1: 840 (1901); Engl. in E.J. 51: 435, fig. 2 (1914); Hutch in F.T.A. 6(2): 217 (1917); Hauman in F.C.B. 1: 96 (1948); C.C. Berg in B.J.B.B. 47: 294, fig. 5 (1977). Type: Zaire, Kimuenzam, *Gillet* 1742 (BR, holo.!)

Shrub to 6 m. or tree up to 35 m. tall. Lamina subcoriaceous to chartaceous, oblong to subobovate or oblanceolate, sometimes elliptic or lanceolate, 2–4(–21) × 1.5–6(–12.5) cm., apex acuminate, base acute to obtuse, margin entire; upper surface glabrous, lower surface puberulous; lateral veins 5–6 pairs, tertiary venation reticulate; petiole 0.3–1(–1.5) cm. long; stipules 0.3–0.7 cm. long, caducous. Staminate inflorescences 3–6 mm. in diameter, subsessile or on peduncle up to 2 mm. long. Bisexual inflorescences 5–10 mm. in diameter, subsessile or with a peduncle up to 4 mm. long; stigmas 2–4.5 mm. long. Infructescences subglobose to ellipsoid, ± 2 cm. in diameter. Fig. 9, p. 18.

TANZANIA. Lindi District: Ngongo, 12 Dec. 1942, *Gillman* 1176! & W. slopes of Rondo Plateau, 21 Nov. 1966, *Gillett* 17956! & 3.6 km. on Nyengedi to Rondo Forest Station road, 20 Oct. 1978, *Magogo & Rose Innes* RR1 387!

DISTR. **T** 6, 8; Congo Brazzaville, Zaire, then disjunctly to SE. Tanzania and Mozambique

HAB. Deciduous thicket; up to 450 m.

SYN. *B. parvifolia* Engl. in E.J. 51: 487, fig. 3 (1914); Hutch. in F.T.A. 6(2): 216 (1917); Peter, F.D.O.-A. 2: 81 (1932); T.T.C.L.: 350 (1949). Type: Tanzania, 'Amani' (? cultivated or mislabelled), *Koerner* in *Herb. Amani* 2259 (B, holo.!)

Trymatococcus parvifolius Engl., V.E. 3(1): 27, fig. 15 (1915), without description, figure based on *Koerner* in *Herb. Amani* 2259 (type of *B. parvifolia*)

NOTE. The populations from the Congo basin occur in primary and old secondary forests on the trees grow there to 35 m., whereas in the east coast thickets the shrubs or treelets only reach 6 m. or so, but no other differences are evident.

9. TRILEPISIUM

Thouars, Gen. Nov. Madag.: 22 (1806); C.C. Berg in B.J.B.B. 47: 297 (1977)

Bosqueia Baillon in Adansonia 3: 338 (1863)

Pontya A. Chev. in Mém. Soc. Bot. Fr. 8: 210 (1912)

Trees, monoecious. Leaves distichous, pinnately veined to sometimes subtriplinerved; stipules fully amplexicaul, connate. Inflorescences in the leaf-axils, initially enveloped by 2 coriaceous bud-scales, bisexual, involucrate. Staminate flowers peripheral, without perianth, initially covered by the membranaceous expanded margin of the receptacle, this cover tearing at anthesis, leaving a tubular part around the pistillate flower and a fringe-like marginal part; stamens without distinct floral arrangement, straight before anthesis. Pistillate flower 1, central, partly adnate to the receptacle; perianth 4-lobed;ovary adnate to the perianth; stigmas 2, band-shaped, equal. Fruit forming a drupaceous whole with the enlarged fleshy dark purple to red receptacle, crowned with the remnants of the stamens and the marginal part of the receptacle. Seed large, without endosperm; cotyledons equal, thick and fused.

One species in tropical Africa.

NOTE. *Bosqueia spinosa* Engl., described from **T** 1, Ukerewe I., was based on a specimen of *Chaetacme aristata* Planch. (Ulmaceae).

T. madagascariensis *DC.*, Prodr. 2: 639 (1825); C.C. Berg in B.J.B.B. 47: 299, figs. 6, 7 (1977) & in Fl. Cameroun 28: 103, t. 35 (1985). Type: Madagascar, *Du Petit-Thouars* (P, holo.!)

Tree up to 25(–40) m. tall, Lamina coriaceous or subcoriaceous, elliptic to oblong or obovate, sometimes lanceolate or oblanceolate, 2–12(–18) × 1.5–6.5(–8) cm., apex acuminate, base acute to obtuse, margin entire, glabrous; lateral veins 4–10(–12) pairs, tertiary venation reticulate; petiole 0.3–1.5 cm. long; stipules 0.2–1.2 cm. long, glabrous, caducous. Inflorescences 0.5–0.8(–1) cm. in diameter; peduncle 0.2–1.5(–2.3 in fruit) cm. long. Filaments 2–10 mm. long; anthers 0.2–0.8 mm. long. Stigmas 2–8 mm. long. Infructescence ovoid to ellipsoid, 1.2–1.8 cm. in diameter. Fig. 10, p. 19.

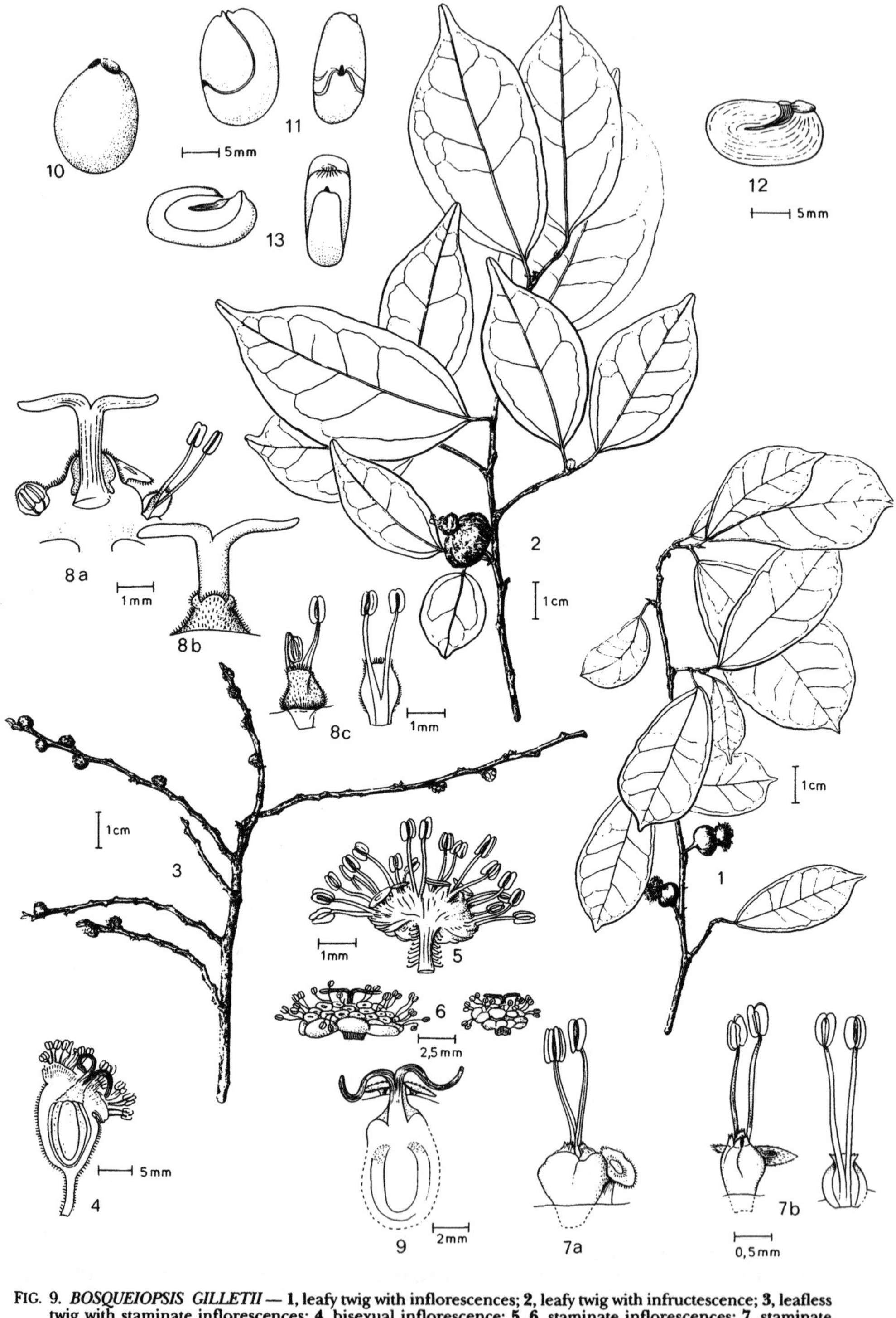

FIG. 9. *BOSQUEIOPSIS GILLETII* — **1**, leafy twig with inflorescences; **2**, leafy twig with infructescence; **3**, leafless twig with staminate inflorescences; **4**, bisexual inflorescence; **5**, **6**, staminate inflorescences; **7**, staminate flowers and bracts; **8**, abortive pistillate flower and staminate flowers; **9**, pistillate flower; **10**, seed, **11**, embryo; **12**, seed, **13**, embryo from another plant. 2, from *Carlier* 272; 3, from *Schlieben* 5437; 4, 5, 7, 9, from *Gillett* 17956; 6, 8, from *Collin* 17; 10, 11, from *Torre & Paiva* 10051; 1, 12, 13, collector not recorded.

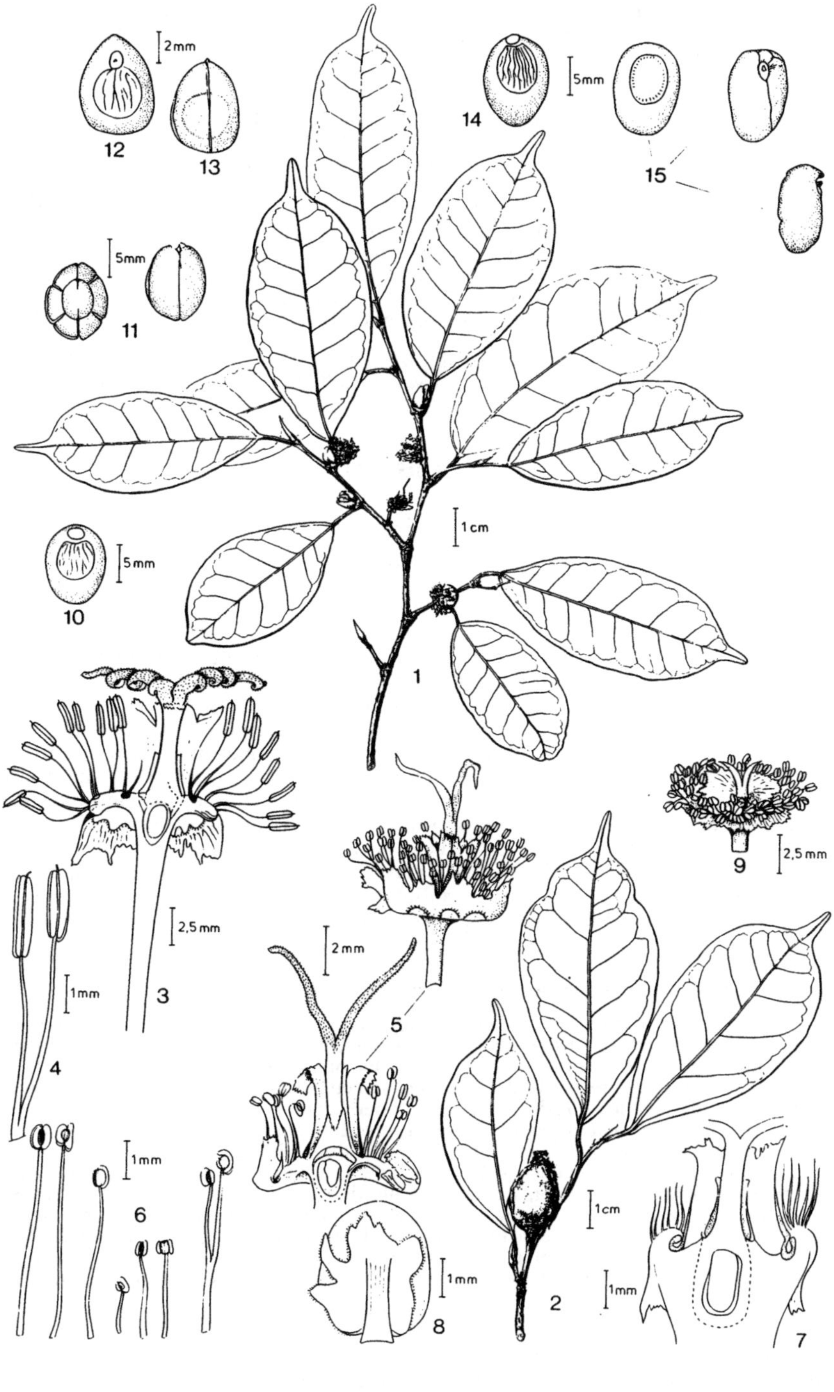

FIG. 10. *TRILEPISIUM MADAGASCARIENSE* — **1**, leafy twig with inflorescences; **2**, leafy twig with infructescence; **3**, inflorescence; **4**, stamens; **5**, inflorescence; **6**, stamens from another plant; **7**, inflorescence; **8**, involucre; **9**, inflorescence; **10**, seed; **11**, embryo; **12**, seed; **13**, embryo showing variation; **14**, seed; **15**, embryo. 1, 2, 10, 11, from *Breteler* 2732; 3, 4, from *Gossweiler* 4394; 5, 6, from *Simão* 506; 7, 8, from *Gossweiler* 6514; 9, from *Mendonça* 141; 12, 13, from *Capuron* 948; 14, 15, from *Devred* 2454. Drawn by E.H. Hupkens van der Elst and W. Scheepmaker.

UGANDA. Bunyoro District: Budongo Forest, Dec. 1934, *Eggeling* 1469!; Kigezi District: Ishasha Gorge, Nov. 1946, *Purseglove* 2267!; Mengo District: Kirerema and Kipayo, 20 Aug. 1913, *Dummer* 197!

KENYA. Nandi, 1929, *Wye* in *F.D.* 1843!; N. Kavirondo District: Kakamega Forest, 10 Dec. 1956, *Verdcourt* 1708!; Kwale, 1919, *R.M. Graham* in *F.D.* 1643!

TANZANIA. Moshi District: Lyamungu, 2 Nov. 1954, *Matalu* 3218!; Lushoto District: Mombo, 21 Dec. 1955, *Muze* 12!; Morogoro District: Turiani, Manyanga Forest, Nov. 1953, *Semsei* 1415!

DISTR. **U** 2, 4; **K** 3–5, 7; **T** 1–3, 6–8; **Z;** Guinée to S. Ethiopia and south to Angola and South Africa (Natal), also Madagascar and Seychelles

HAB. Rain-forest and other wetter evergreen forests, riverine and ground-water forest; 1800 m.

SYN. *Bosqueia phoberos* Baillon in Adansonia 3: 339 (1863); Hutch. in F.T.A. 6(2): 219 (1917); T.T.C.L.: 350 (1949); I.T.U., ed. 2: 234 (1952); K.T.S.: 309 (1961); Hamilton, Ug. For. Trees: 96 (1981). Type: Tanzania, Zanzibar, *Boivin* (P, holo.!)

B. angolensis Ficalho, Pl. Ut. Afr. Port.: 27 (1884); Hutch. in F.T.A. 6(2): 218 (1917); Hauman in F.C.B. 1: 95 (1948). Type: Angola, Golungo Alto, *Welwitsch* 456 (LISU, holo.!, B, BM, G, K, P, iso.!)

B. cerasifolia Engl., E.M. 1: 36 (1898) & in E.J. 51: 439, fig. 5F–H (1914); Hutch. in F.T.A. 6(2): 219 (1917). Type: Tanzania, Kilimanjaro, *Volkens* 1935 (B, holo.!, BM, BR, G, iso.!)

10. DORSTENIA

L., Sp. Pl.: 121 (1753) & Gen. Pl., ed. 5: 56 (1754)

Kosaria Forssk., Fl. Aegypt.-Arab.: 164 (1775)

Craterogyne Lanjouw in Rec. Trav. Bot. Néerl. 32: 272 (1935)

Herbs, often at least somewhat succulent, rhizomatous or tuberous, or shrubs. Leaves in spirals or distichous, pinnately, less often palmately or radiately veined; stipules lateral, free. Inflorescences bisexual, in the leaf-axils, discoid to turbinate, in outline circular, elliptic, ± stellate or naviculate, receptacle outside bracteate or mostly with marginal and/or also submarginal appendages. Flowers connate. Staminate flowers numerous; tepals (1–)2–3(–4), free or basally connate; stamens 2–3; pistillode usually absent. Pistillate flowers in the central part of the inflorescence, 1–numerous; perianth tubular, with only the apex free from the surrounding flowers; ovary free; stigmas 2, filiform to band-shaped, equal or unequal, or 1. Fruit a dehiscent drupe, the white fleshy part pushing out the endocarp-body (pyrene) if large or ejecting the endocarp-body if small; endocarp-body (especially if large) subglobose and smooth or tetrahedral and tuberculate. Seed large and without endosperm or small and with endosperm; cotyledons thick and unequal or flat and equal.

About 105 species, ± 45 in the Neotropics, 1 in Asia and ± 58 in Africa.

Beside clear-cut species the genus comprises many taxa showing variation patterns reminiscent of that found in apomictic (agamospermous) or autogamous temperate species or species complexes. Their delimitation is therefore not always quite satisfactory. In several species more or less distinct morphological entities are recognised as varieties, sometimes as informal entities. Extensive field-work and knowledge of the reproductive biology is needed to improve the taxonomy of the genus.

The shape of the receptacle varies widely from narrowly boat-shaped (naviculate), with a terminal appendage at either end, to stellate or circular in outline with appendages or bracts variously inserted on the margin or on the back of the receptacle. The extremes are very different and provide useful criteria for identification, but, in using the following key, it should be appreciated that several species vary across this division, though even then variants or populations tend to polarise to a limited extent one way or the other. The number and size of appendages and the extent to which they are marginal, submarginal or in two distinct rows is also somewhat labile, so that trial of alternative leads in the couplets of the key may be advisable with some specimens of the common variable species.

Stems woody, forming small shrubs:
- Receptacle discoid to broadly turbinate, with small bracts around the edge in 2 rows. . . . . 1. *D. kameruniana*
- Receptacle boat-shaped, with long terminal appendages at either end . . . . 2. *D. alta*

Stems herbaceous or succulent, occasionally becoming a little woody below, but not forming shrubs:

Leaves crowded to rosulate; stems succulent and thick or very short from a tuber, internodes remaining short:
  Receptacle elongate . . . . 23. *D. barnimiana*
  Receptacle isodiametric:
    Leafy stem up to 15 cm. long; petiole 0.1–2.5 cm. long; peduncle 1–3 cm. long . . . . 22. *D. foetida*
    Leafy stem up to 1 cm. long from a tuber; petiole 1–8 cm. long; peduncle 6–14.5 cm. long . . . . . . 24. *D. ellenbeckiana*
Leaves well spaced at least below the apex of the stem:
  Appendages in a single row from the margin of the receptacle (even if only 2 and even (*D. thikaensis*) if margin bent, forming a step down to the base of the appendage):
    Face of receptacle ± isodiametric with appendages of varying length:
      Stems sappy to slightly succulent, trailing and ascending, early glabrescent; petiole puberulous in adaxial channel, glabrous on the back; stipules caducous; receptacle with green margin 1–3 mm. wide, with 1–4 appendages longer than the others . . . . 14. *D. goetzei*
      Stems herbaceous to slightly fleshy; petiole hairy all round, sometimes minutely; stipules ± persistent; receptacle, if margined, with more numerous or less disparate longer appendages:
        Margin of the receptacle 3–5 mm. wide, with conspicuous radiating stripes, with many tooth-shaped appendages to 4 mm. long (terminal ones to 7 mm.) . . . . 9. *D. schliebenii*
        Margin of the receptacle 0–1.5 mm. wide, without stripes:
          Plants robust, erect, (20–)30–90(–120) cm. tall, minutely puberulous on youngest parts; lamina (4–)6–23(–29) × (2–)4–10 cm., glabrous or slightly puberulous beneath; receptacle subcircular to multiangular or substellate, with ± 40 appendages 1–40 mm. long . . . . 10. *D. holstii*
          Plants creeping and ascending, often rooting at nodes, some stems erect but slender and hairy, 10–50 cm. tall; lamina (1–)3–19 × (0.5–)1–6 cm., often conspicuously hairy or scabrid:
            Terminal appendages of receptacle 7–45 mm. long . . . . 3. *D. tayloriana*
            Terminal appendages of receptacle up to 5 mm. long:
              Leaves scabrid above, coarsely toothed towards apex; receptacle funnel-shaped, densely hispid, with subequal appendages shorter than depth of the receptacle . . . . 11. *D. brownii*
              Leaves generally glabrous above, rarely scabridulous, sinuate to pinnatifid or lobed; receptacle glabrous to minutely puberulous outside, with some lobes at least generally longer than the depth of the receptacle:
                Receptacle-face isodiametric, with a margin 0.3–2(–3) mm. wide with appendages 0.2–4 mm. long, 2–3 slightly longer than the others 4. *D. variifolia*

Receptacle-face generally longer than broad, oblong to elliptic or rhombic, rarely circular, with a margin up to 0.5 mm. wide, with appendages mostly 0.2–0.5 mm. long, the terminal ones up to 5 mm. . . . . . . 6. *D. ulugurensis*

Face of the receptacle elongate, with ± well-developed appendages at either end, with or without shorter lateral appendages:

Lateral appendages absent:

Receptacle held erect, narrowly boat-shaped, with a long terminal appendage pointing upwards, and a shorter one pointing downwards 13. *D. psilurus*

Receptacle not vertical nor with such disparate appendages:

Margin of the receptacle stepped down to the terminal appendages; appendages linear, 1.5–8 mm. long . . . . 7. *D. thikaensis*

Margin of the receptacle extended directly into the appendages; appendages linear to spathulate, 15–27 mm. long . . . . . 8. *D. bicaudata*

Lateral appendages present:

Margin of the receptacle 3–5 mm. wide, with conspicuous radiating stripes . . . . . 9. *D. schliebenii*

Margin of the receptacle up to 1.5 mm. wide, without radiating stripes:

Hairs in longitudinal lines down the stem; receptacle with the linear upper appendage 10–17 mm. long, the lower 3–8 mm. long, and many tiny lateral tooth-like appendages 12. *D. afromontana*

Hairs all round stem; receptacle-appendages less disparate:

Terminal appendages 7–45 mm. long; lateral appendages 10–33, (0.2–)1–17 mm. long 3. *D. tayloriana*

Terminal appendages 1–5 mm. long; lateral appendages 10–20, up to 1(–1.5) mm. long:

Peduncle recurved; plant up to 60 cm. tall; lamina coarsely dentate . . . . . 5. *D. dionga*

Peduncle almost erect; plant up to 30 cm. tall; lamina entire, sinuate or pinnately incised . . . . 6. *D. ulugurensis*

Appendages in 2 rows, the inner row marginal (only a few teeth in *D. warneckei*), the outer row (even if only 2) inserted below the receptacle-margin:

Receptacle held erect, boat-shaped, with 2 long outer appendages, the longer one generally 5–13 cm. long; stem erect or creeping from a tuber . . . . 19. *D. buchananii*

Receptacle elliptic to isodiametric, normally with 3–many outer appendages:

Stems annual, erect, herbaceous, from a discoid to subglobular tuber or series of superposed tubers, with scale leaves on lower part; leaves hairy on both surfaces, generally entire to finely denticulate, sometimes coarsely dentate:

Petiole 0–2(–5) mm. long; lamina usually ± scabrous; receptacle variously shaped but not stellate; stigmas (1–)2 . . . . 20. *D. benguellensis*

Petiole (2–)5–25 mm. long; leaves usually smooth; receptacle stellate or substellate; stigma 1 21. *D. cuspidata*

Stems trailing to erect, sappy or succulent, basal parts sometimes swollen and forming irregular tubers, otherwise rhizomatous, without scale leaves; leaves glabrous or sparsely puberulous generally only on the veins, entire to commonly repand or coarsely to irregularly toothed:

Receptacle-face subcircular, with inner marginal row of appendages up to 8 mm. long, outer row 8–12, filiform, up to 55(–70) mm. long; stipules caducous; stems ascending to erect to 1 m., puberulous . . . . 15. *D. tenuiradiata*

Receptacle with shorter appendages; stipules ± persistent:

Appendages of inner row few, short, of outer row 10–15, triangular to ligulate or subspathulate, 0.5–10 mm. long; stems generally erect to 1 m., not swollen, subglabrous or with hairs in longitudinal lines . . . . 16. *D. warneckei*

Appendages of inner row numerous, of outer row either few or narrower; stems often trailing or swollen basally:

Face of the receptacle narrowly elliptic to sub-stellate or circular, with outer row of appendages 2–12 and generally subulate to filiform; hairs (if present) generally all round stem; scars of leaves, stipules and inflorescences conspicuous and prominent; stigmas 2 . . . . . . 17. *D. hildebrandtii*

Face of the receptacle 3–4-angular, irregularly substellate or less often oval, with outer row of appendages (2–)3–7, triangular or band-shaped (longer at corners); hairs in ± distinct lines down stem; scars of leaves, etc. not conspicuous and prominent; stigma 1 . . . . 18. *D. zanzibarica*

1. D. kameruniana *Engl.* in E.J. 20: 142 (1894); C.C. Berg in Bot. Notis. 131: 62, t. 6 (1978); C.C. Berg et al., Fl. Gabon 26: 34 (1984); Hijman in Fl. Cameroun 28: 32 (1985). Type: Cameroun, Lokoundjé, *Dinklage* 232 (B, holo.!, HBG, iso.!)

Shrub or undershrub, 0.5–3(–6) m. tall. Leafy twigs 0.5–2 mm. thick, puberulous to hirtellous. Leaves (at least on the branches) distichous; lamina subcoriaceous to chartaceous when dry, lanceolate to elliptic or obovate, (3–)7–15(–22) × (1–)2.5–9 cm., apex acuminate to subcaudate, sometimes truncate and coarsely dentate to lobed, base cuneate to obtuse, margin dentate (towards the apex sometimes lobate) to subentire; both surfaces puberulous; lateral veins (4–)6–12(–14) pairs; petiole 0.3–1(–1.7) cm. long, ± 1 mm. thick; stipules narrowly triangular, 3–11 mm. long, subpersistent. Inflorescences solitary or sometimes up to 3 together; peduncle 0.2–0.8(–1.2) cm. long. Receptacle broadly turbinate to discoid, subcircular to circular in outline, 0.3–0.8 cm. in diameter, with ± 2 rows of small reniform to ovate bracts at the margin, some bracts lower. Staminate flowers crowded; perianth-lobes 2, 3–6-lobed or -partite; stamens 2. Pistillate flowers 1(–2), central; perianth tubular; stigmas 2. Infructescences turbinate to subglobose, 0.7–0.8 cm. in diameter; endocarp-body subglobose, ± 6.5 mm. in diameter, smooth. Fig. 11, p.24.

UGANDA. Bunyoro District: Budongo, Feb. 1932, *Rolfe* 438! & Feb. 1939, *Eggeling* 3827!; Mengo District: Mulange, Sept. 1919, *Dummer* 4264!

KENYA. Kwale District: Gongoni [Gogoni] Forest, Aug. 1936, *Dale* in *F.D.* 3545! & Shimba Hills, Mwele Mdogo Forest, 6 Feb. 1953, *Drummond & Hemsley* 1154!; Kilifi District: Cha Simba, 30 Dec. 1970, *Faden et al.* 70/947!

TANZANIA. Lushoto District: Mangubu–Misoswe, 14 Feb. 1931, *Greenway* 2894!; Morogoro District: Turiani Falls, 4 Nov. 1947, *Brenan & Greenway* 8275!; Iringa District: Sanje Falls, 23 July 1983, *Polhill & Lovett* 5119!

DISTR. **U** 2, 4; **K** 7; **T** 3, 6, 7; extending to Angola and Cameroun, also in W. Africa (Ghana to Guinée)

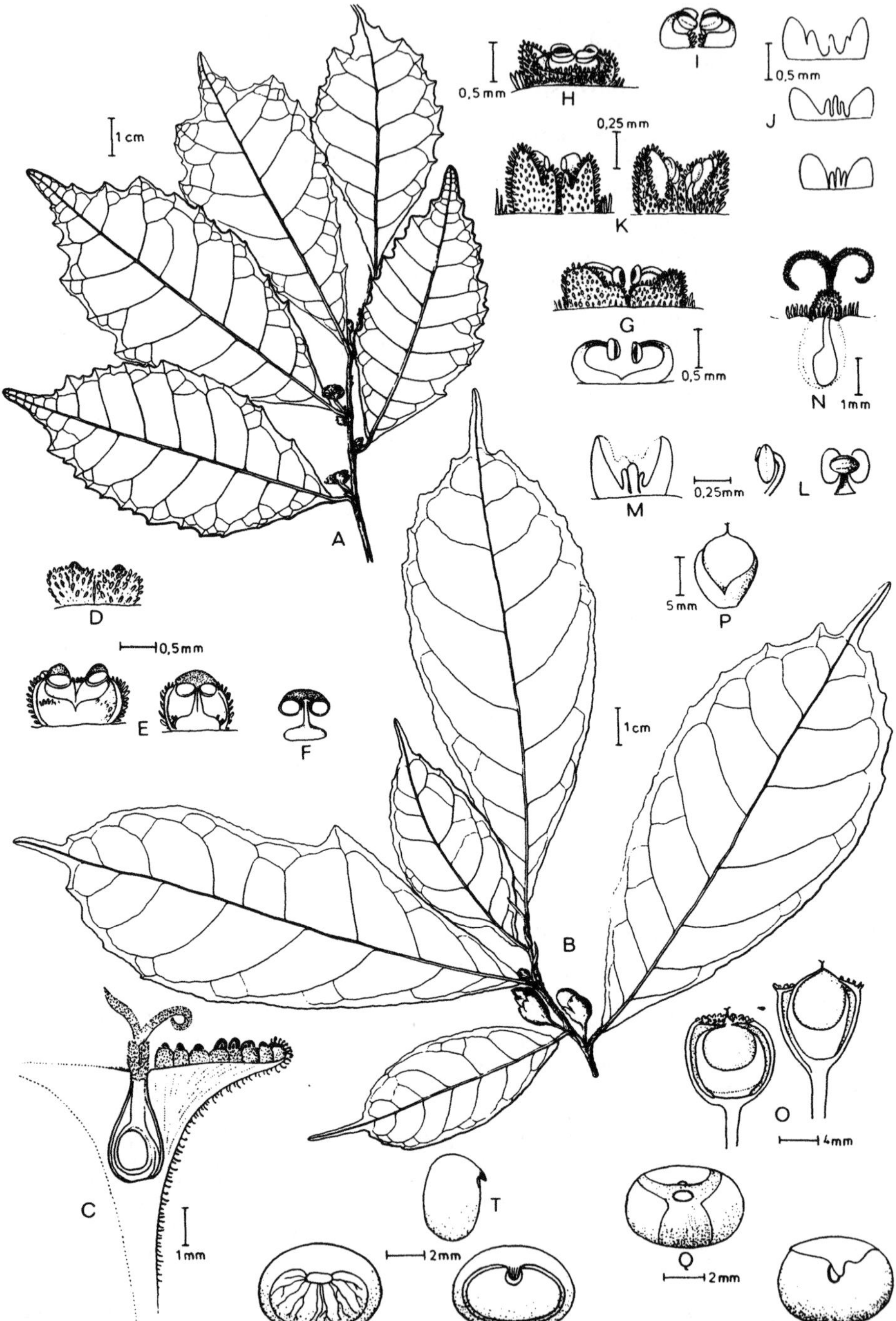

FIG. 11. *DORSTENIA KAMERUNIANA* — **A**, leafy twig with inflorescences; **B**, leafy twig with infructescences; **C**, inflorescence; **D**, staminate flower; **E, F**, stamens; **G**, staminate flower and stamens from another plant; **H, I**, same, showing variation; **J**, tepals; **K, L**, staminate flower and stamens; **M**, tepals; **N**, pistillate flower; **O**, infructescence; **P**, fruit; **Q**, seed; **R**, embryo; **S, T**, another seed and embryo. A, from *Procter* 2786; B, from *Leeuwenberg* 3886; C, from *de Wilde* 1638; D–F, from *Leeuwenberg* 5184; G, S, T, from *Callens* 3738; H–J, from *Gilbert* 2090; K–M, from *Brenan & Greenway* 8275; N, from *Evrard* 5736; O, from *Tisserant* 998; P–R, from *Breteler* 1382. Drawn by E.H. Hupkens van der Elst and W. Scheepmaker.

HAB. Undergrowth of evergreen forests, often near streams, sometimes in secondary growth; up to 1300 m.

SYN. *Trymatococcus kamerunianus* (Engl.) Engl., E.M. 1: 29, t. 11B (1898); Rendle in F.T.A. 6(2): 76 (1916); I.T.U., ed. 2: 237 (1952)
T. usambarensis Engl. in E.J. 33: 117 (1902); Peter, F.D.O.-A. 2: 81 (1932). Type: Tanzania, Pangani District, Makinyumbe [Makingumbi], *Scheffler* 251 (B, holo.!)
Craterogyne kameruniana (Engl.) Lanjouw in Rec. Trav. Bot. Néerl. 32: 274, t. 7, 8 (1935); Hauman in F.C.B. 1: 81 (1948); T.T.C.L.: 351 (1949); F.W.T.A., ed. 2, 1: 599 (1958); K.T.S.: 311 (1961)

2. D. alta *Engl.* in E.J. 40: 545 (1908); Rendle in F.T.A. 6(2): 59 (1916); Peter, F.D.O.-A. 2: 77 (1932); T.T.C.L.: 352 (1949). Type: Tanzania, Lushoto District, Amani, *Warnecke* 310 (B, holo.!, BM, K, iso.!)

Shrub up to 2 m. tall. Stems branched, puberulous. Leaves distichous; lamina chartaceous when dry, elliptic to oblong or obovate, (3–)5–17(–20) × (1–)2–8 cm., apex acute to acuminate, base cuneate to subobtuse, margin entire or sublobate; upper surface glabrous or puberulous on the midrib, lower surface glabrous or sparsely puberulous; lateral veins 5–8(–14) pairs; petiole 2.5–6(–9) cm. long, 0.5–1 mm. thick; stipules triangular to subulate, 1–6 mm. long, subpersistent. Inflorescences solitary; peduncle 0.4–1.4 cm. long, 0.5–1 mm. thick. Receptacle naviculate, 1.5–4 × 0.5–1 cm.; flowering face linear, margin 3–8 mm. wide, undulate; appendages 2, terminal, linear, (2.5–)7–30 mm. long, sometimes with a third tooth-like appendage laterally. Staminate flowers ± 45, ± spaced; perianth-lobes 1–3; stamens 1–3. Pistillate flowers 1–2; perianth disciform; stigmas 2. Endocarp-body subglobose, 5–7 mm. in diameter, smooth.

KENYA. Kwale District: Shimba Hills, 11 Feb. 1953, *Drummond & Hemsley* 1199!; Kilifi District: Cha Simba, 30 Dec. 1970, *Faden et al.* 70/944! & 17 Nov. 1974, *B. Adams* 103!
TANZANIA. Tanga District: E. Usambara Mts., between Ngua and Magunga estates, 17 July 1953, *Drummond & Hemsley* 3351!; Morogoro District: Mtibwa, Nov. 1953, *Paulo* 174! & 177!; Ulanga District: Itula, July 1960, *Haerdi* 572/0!
DISTR. **K** 7; **T** 3, 6; Congo (Brazzaville)
HAB. Undershrub in evergreen forests, often near streams; up to 800 m.

SYN. *D. orientalis* De Wild., Pl. Bequaert. 6: 49 (1932); T.T.C.L.: 352 (1949). Type: Tanzania, Lushoto District, Amani, *Sacleux* 2318 (P, holo.!)

3. D. tayloriana *Rendle* in J.B. 53: 300 (1915) & in F.T.A. 6(2): 43 (1916); Peter, F.D.O.-A. 2: 76 (1932). Type: Kenya, Kilifi District, Rabai Hills, Mtoni stream, *W.E. Taylor* (BM, holo.!)

Herb up to 50 cm. tall, rhizomatous; stems ascending, unbranched, 1.5–4 mm. thick, whitish to sometimes yellowish puberulous or hirtellous, the lower part woody. Leaves in spirals, crowded at the apex of the stem; lamina chartaceous when dry, oblanceolate to obovate, (1–)3–14 × (0.5–)1–6 cm., apex acute to obtuse, base obtuse to subcordate or occasionally cuneate, margin repand to faintly or sometimes coarsely dentate; upper surface smooth or ± scabrous, lower surface scabrous or smooth and puberulous to hirtellous on the main veins; lateral veins 3–8(–10) pairs; petiole 0.2–1.3 cm. long, 0.5–1.5 mm. thick; stipules either triangular, ± 0.1 mm. long and caducous, or subulate, 4–10 mm. long and persistent. Inflorescences solitary; peduncle 0.6–2.7 cm. long, gradually passing into the receptacle. Receptacle zygomorphic to almost actinomorphic, naviculate to broadly funnel-shaped and orbicular in outline; flowering face linear, elliptic or subrhombic to subcircular, 0.5–2.2 × 0.4–1 cm., plane, often dark purplish, margin 0.1–1.5 mm. wide, often with a ridge bordering the flowering face; primary appendages 2, terminal (in actinomorphic inflorescences hardly distinct from secondary appendages), linear, 7–45 mm. long, secondary appendages 10–33, lateral, triangular to linear, (0.2–)1–17 mm. long. Staminate flowers ± spaced; perianth-lobes 2–3; stamens 2–3. Pistillate flowers up to ± 10; perianth tubular; stigmas 2. Endocarp-body tetrahedral, ± 2.5 mm. in diameter, two sides tuberculate.

var. **tayloriana**

Herb up to 50 cm. tall. Upper surface of the lamina often ± scabrous; stipules narrowly triangular, ± 1 mm. long, caducous. Receptacle zygomorphic, naviculate, terminal appendages distinctly longer than the secondary appendages; flowering face elliptic, irregularly rhombic or sometimes linear.

KENYA. Kilifi District: Rabai Hills, Mtoni stream, Sept. 1885, *W.E. Taylor*!

TANZANIA. Lushoto District: Lower Sigi valley, 15 May 1950, *Verdcourt et al.* 199!; Tanga District: Amboni Caves, 1 Aug. 1953, *Drummond & Hemsley* 3591!; Pangani District: Bushiri Estate, 4 May 1950, *Faulkner* 565!

DISTR. **K** 7; **T** 3; not known elsewhere

HAB. Lowland evergreen forest, often on limestone; up to 600 m.

SYN. *D. amboniensis* De Wild., Pl. Bequaert. 6: 19 (1932); T.T.C.L.: 352 (1949). Type: Tanzania, Tanga District, Amboni Forest, *Sacleux* 2317 (BR, holo.!, P, iso.!)

var. **laikipiensis** (*Rendle*) *Hijman*, stat. nov. Type: Kenya, Laikipia Plains, *Battiscombe* 61 (K, holo.!)

Herb up to 30 cm. tall. Both surfaces of the lamina sometimes scabrous; stipules subulate, 4–10 mm. long, persistent. Receptacle tending to actinomorphic, broadly naviculate to funnel-shaped, terminal appendages hardly longer than the secondary appendages; flowering face elliptic to subcircular.

KENYA. Kwale District: Gongoni Forest, *Gardner* in *F.D.* 1440! & Buda Mafisini Forest, 22 Aug. 1953, *Drummond & Hemsley* 3959!; Kilifi District: Chonyi–Ribe road, 1 km. NE. of Pangani, 28 July 1974, *Faden* 74/1271!

TANZANIA. Morogoro District: Turiani, Manyangu Forest, Nov. 1953, *Semsei* 1420! & E. Uluguru Mts., Mkungwe, 22 May 1933, *Schlieben* 3969! & Shikurufumi [Chigurufumi] Forest, Mar. 1955, *Semsei* 2035!

DISTR. **K** 3, 7; **T** 3, 6; Mozambique

HAB. Evergreen forest, principally coastal, but up 1950 m.

SYN. *D. laikipiensis* Rendle in J.B. 53: 299 (1915) & in F.T.A. 6(2): 34 (1916)
D. pectinata Peter, F.D.O.-A. 2: 74, Descr. 5, t. 6/1a–c (1932). Type: Tanzania, Tanga District, near Amboni, Ukereni Hill, *Peter* 25589 (B, holo.!)
D. rugosa Peter, F.D.O.-A. 2: 74, Descr. 5, t. 5/2a–c (1932). Type: Tanzania, E. Usambara Mts., Magunga–Kitiwu, *Peter* 12830 (B, holo.!)

NOTE. The type specimen may be mislabelled as it is the only record from inland at high altitudes.

4. D. variifolia *Engl.* in E.J. 28: 376 (1900); Rendle in F.T.A. 6(2): 35 (1916); Peter, F.D.O.-A. 2: 74 (1932). Type: Tanzania, Iringa District, Uzungwa [Utschungwa] Mts., near Pongelo, *Goetze* 613 (B, holo.!)

Herb 10–50 cm. tall, rhizomatous; rhizomes ± woody; stems ascending, sometimes branched, 1–3 mm. thick, puberulous. Leaves in spirals; lamina chartaceous when dry, obovate to oblanceolate, sometimes ovate, oblong or linear, (1–)3–19 × (0.4–)1–6 cm., apex acute to subobtuse, base obtuse to cuneate, margin repand to sinuate, pinnatifid to -lobed, with 2–8 lobes at each side, coarsely dentate or irregularly pinnately incised; upper surface glabrous and smooth or scabridulous, lower surface puberulous mainly on the veins; lateral veins 2–8 pairs; petiole 0.2–0.9 cm. long, ± 1 mm. thick; stipules subulate, 1.5–5 mm. long, persistent. Inflorescences solitary or in pairs; peduncle 1–3.5 cm. long, ± 0.5 mm. thick. Receptacle slightly funnel-shaped, 0.3–1 cm. in diameter; flowering face subcircular, 0.2–0.7 cm. in diameter, plane, margin 0.3–2(–3) mm. wide; appendages in 1 row, 15–28, linear or triangular, 0.2–4 mm. long (2 or 3 slightly longer than the others). Staminate flowers spaced; perianth-lobes 3; stamens 3. Pistillate flowers 4–12; perianth conical; stigmas 2. Endocarp-body slightly tetrahedral, ± 3 mm. in diameter, 2 sides coarsely tuberculate.

TANZANIA. Ulanga District: Mahenge Plateau, Issongo, 21 Apr. 1932, *Schlieben* 2113! & Sali, Ngongo Mt., 22 Jan. 1979, *Cribb et al.* 11121!; Iringa District: Uzungwa Scarp Forest Reserve, 29 Jan. 1971, *Mabberley* 622!

DISTR. **T** 6, 7; not known elsewhere

HAB. Forest, in moist shady places; 900–1850 m.

5. D. dionga *Engl.* in E.J. 28: 377 (1900); Rendle in F.T.A. 6(2): 46 (1916); Peter, F.D.O.-A. 2: 76 (1932). Type: Tanzania, S. Uluguru Mts., *Goetze* 177 (B, holo.!)

Herb up to 80 cm. tall, rhizomatous; stems erect, unbranched, 3–5 mm. thick, puberulous. Leaves in spirals, tending to distichous; lamina chartaceous when dry, elliptic to obovate. (3.5–)7–17 × 3.5–8.5 cm., apex acute to acuminate, base cuneate to subobtuse, margin coarsely dentate; upper surface glabrous, lower surface sparsely puberulous; lateral veins 6–8 pairs; petiole 1–2.6 cm. long, 1–2.5 mm. thick; stipules subulate, 2.5–5.5 mm. long, subpersistent. Inflorescences (1–)2(–3) together; peduncle 0.5–1 cm. long, ± 1 mm. thick, recurved. Receptacle funnel-shaped, laterally compressed, 0.6–1 cm. long; flowering face narrowly elliptic, 0.5–0.8 × 0.2 cm., margin ± 0.5 mm. wide;

appendages in 1 row, primary ones 2, terminal, obovate, 1–5 mm. long, 1 mm. wide, concave, secondary ones 10–16, lateral, obovate to ovate or subtriangular, 0.3–1 mm. long, concave. Staminate flowers ± spaced; perianth-lobes 3; stamens 3. Pistillate flowers up to ± 10; perianth conical; stigmas 2, linear.

TANZANIA. Morogoro District: Uluguru Mts., Mgeta R., below Hululu Falls, 15 Mar. 1953, *Drummond & Hemsley* 1581! & S. Uluguru Mts., Nov. 1899, *Goetze* 177!
DISTR. **T** 6; not known elsewhere
HAB. Moist and shady places in forest; 1200–1800 m.

NOTE. This species resembles *D. ulugurensis* and may not be distinct (see note under that species). The specific epithet derives from the vernacular name 'dionga'.

6. D. ulugurensis Engl., E.M. 1: 13, t. 5B (1898) & in E.J. 28, 376 (1900) & V.E. 3(1): 23 (1915); Rendle in F.T.A. 6(2): 43 (1916); Peter, F.D.O.-A. 2: 76 (1932). Type: Tanzania, Uluguru Mts., Nghweme [Nghwenn], *Stuhlmann* 8800 (B, holo.!)

Herb 15–30 cm. tall, rhizomatous; stems ascending to creeping, unbranched, 1–3 mm. thick, puberulous. Leaves almost distichous; lamina chartaceous when dry, oblong to elliptic, 3–16 × 2–8 cm., apex acute, base cuneate to obtuse, pinnatifid to -lobed, with up to 15 lobes on each side; upper surface glabrous and smooth or scabridulous, lower surface puberulous; lateral veins 5–9 pairs; petiole 0.5–2 cm. long, ± 1 mm. thick; stipules subulate, 1.5–7.5 mm. long, persistent. Inflorescences 1–2(–3) together; peduncle 0.5–1.5 cm. long, ± 0.5 mm. thick, almost erect. Receptacle funnel-shaped and laterally compressed to naviculate, 0.2–1 × 0.2–0.5 cm.; flowering face oblong to elliptic or rhombic to subcircular, 0.2–0.8 × 0.2–0.4 cm., margin up to 0.5 mm. wide; appendages in 1 row, primary ones 2, terminal, subobovate to subspathulate, 1–5 mm. long, secondary ones up to ± 20, lateral, triangular to ovate, 0.2–0.5 mm. long, sometimes only 1 or 2 and then up to 1.5 mm. long and at the middle of the receptacle. Staminate flowers ± spaced; perianth-lobes (2–)3; stems (2–)3. Pistillate flowers up to ± 10; perianth conical; stigmas 2. Endocarp-body slightly tetrahedral, ± 2 mm. in diameter, 2 sides tuberculate.

TANZANIA. Morogoro District: Uluguru Mts., Matombo road, Tanana, Feb. 1935, *E.M. Bruce* 809! & Lukwangule Plateau, 24 Feb. 1931, *Schlieben* 3570! & Nghweme [Nghwenn], 18 Oct. 1894, *Stuhlmann* 8800!
DISTR. **T** 6; not known elsewhere
HAB. Montane forest; 1600–2000 m.

NOTE. This species is morphologically similar to *D. dionga*, but the latter is a larger herb, with thicker stems and larger leaves; the leaves of *D. dionga* are not pinnatifid, but coarsely dentate; the inflorescences of *D. dionga* show a shorter and thicker peduncle and the receptacles are more naviculate. The two elements might prove to be conspecific when better known.

7. D. thikaensis *Hijman*, sp. nov. a *D. bicaudata* Peter appendicibus linearibus 1.5–8 mm. longis subter margine insertis facile distinguiter. Type: Kenya, Fort Hall District, Thika, Thika R., *Faden* 66/243 (EA, holo.!, K, iso.!)

Herb up to 50 cm. tall, rhizomatous; rhizome with brownish filiform branched roots up to 0.5 mm. thick, as well as with yellowish fleshy unbranched roots up to ± 2.5 mm. thick; stems procumbent or ascending, unbranched, ± 2.5 mm. thick, puberulous to hirtellous. Leaves in spirals; lamina subcoriaceous to chartaceous when dry, elliptic to oblong, oblanceolate or linear, (3–)6–11 × 1–2.5 cm., apex acute, base cuneate to subobtuse; margin repand to dentate; upper surface scabrous, often with paler spots along the midrib and lateral veins, lower surface puberulous to hispidulous, often with purplish spots along the midrib and lateral veins; lateral veins 4–7 pairs; petiole 0.7–1.2 cm. long, 1–2 mm. thick; stipules narrowly triangular, ± 1 mm. long, persistent. Inflorescences solitary; peduncle up to 1.3 cm. long, 0.5–1 mm. thick, gradually passing into the receptacle. Receptacle naviculate, 1–2.1 × 0.3–0.4 cm.; flowering face linear, concave to plane, 0.8–1.5 × ± 0.25 cm., margin ± 0.5 mm. wide, entire; appendages 2, terminal, inserted below the face on a downwardly bent margin, linear, 1.5–8 mm. long. Staminate flowers ± crowded; perianth-lobes 3; stamens 3. Pistillate flowers 5–7, ± in a median row: perianth conical; stigmas 2. Endocarp-body slightly tetrahedral, ± 2.5 mm. in diameter, slightly tuberculate.

KENYA. S. Nyeri District: Mwea R., *Battiscombe* 689a!; Fort Hall District: Thika Falls, 2 Mar. 1968, *Faden* 68/002! & 12 Apr. 1968, *Faden* 68/137!
DISTR. **K** 4; not known elsewhere

HAB. Riverine forest; 1200–1450 m.

SYN. [*D. scaphigera* sensu Agnew, U.K.W.F.: 318 (1974), *non* Bureau]

NOTE. The remarkable swollen roots have not been found in other species of *Dorstenia*.

8. D. bicaudata *Peter*, F.D.O.-A. 2: 76, Descr. 6, t. 7a, b (1932). Type: Tanzania, Lushoto District, near Amani, Kwamkuyu Falls, *Peter* 17080 (B, lecto.!)

Herb up to 25 cm. tall, rhizomatous; stems creeping or ascending, unbranched, 1–2 mm. thick, puberulous. Leaves in spirals; lamina chartaceous when dry, elliptic, (1.5)3–9.5 × (1–)2–4 cm., apex acute to obtuse, base cuneate to subobtuse, margin 3–5-lobed to sinuate or faintly dentate; upper surface glabrous, lower surface puberulous; lateral veins 3–5 pairs; petiole 0.7–1.5 cm. long, ± 1 mm. thick; stipules narrowly triangular, 0.5–1.5 mm. long, persistent. Inflorescences solitary or in pairs; peduncle 0.4–1.5 cm. long, 0.5–1 mm. thick, gradually passing into the receptacle. Receptacle naviculate; flowering face narrowly elliptic, 0.8–1.3 × ± 0.3 cm., sometimes slightly angular, margin 0.2–0.3 mm. wide; appendages in 1 row, primary ones 2, terminal, linear, 15–27 mm. long. Staminate flowers spaced; perianth-lobes 3; stamens 3. Pistillate flowers 3–8; perianth tubular; stigmas 2. Endocarp-body subglobose, ± 3 mm. in diameter, tuberculate.

TANZANIA. Lushoto District: Derema–Ngambo, 28 May 1950, *Verdcourt* 228! & Dodwe–Kiumba, 2 July 1906, *Zimmerman* 7190! & Kwamkuyu Falls, 12 June 1916, *Peter* 17080!
DISTR. T 3; not known elsewhere
HAB. Rain-forest; 800–900 m.

SYN. *D. bicaudata* Peter var. *quercifolia* Peter, F.D.O.-A. 2: 77, Descr. 6 (1932). Type: Tanzania, Lushoto District, Kwamkuyu Falls, *Peter* 21801b (B, holo., not seen, probably destroyed)

NOTE. *D. bicaudata* resembles *D. afromontana* and *D. thikaensis*.

9. D. schliebenii *Mildbr.* in N.B.G.B. 11: 396 (1932). Type: Tanzania, Ulanga District, Masagati, *Schlieben* 1109 (B, holo.†, G, iso.!)

Herb up to 1 m. tall, rhizomatous; stems erect or ascending, unbranched, up to ± 5 mm. thick, minutely puberulous to yellowish hirtellous, the lower part woody. Leaves in spirals; lamina chartaceous to subcoriaceous when dry, subobovate or obovate, 6–15 × 2–5 cm., apex acute to obtuse, base cuneate to obtuse or cordate, margin entire, sometimes coarsely dentate; upper surface scabrous, lower surface scabrous, partly puberulous (mainly on the veins) or minutely puberulous to yellowish hirtellous; venation impressed and conspicuous above, lateral veins 6–10 pairs; petiole 0.5–1.5 cm. long, 1–3 mm. thick; stipules narrowly triangular to subulate, 2–4.5 mm. long, subpersistent. Inflorescences solitary or sometimes in pairs; peduncle 0.5–1.2 cm. long. Receptacle slightly to distinctly zygomorphic, discoid and plane, elliptic to subcircular to multiangular, 2–3 × 1.2–1.5 cm., margin 3–5 mm. wide, with conspicuous radiating stripes; appendages in 1 row, primary ones 2, terminal, linear to spathulate, 5–7 mm. long, secondary ones numerous, lateral, tooth-shaped, up to 4 mm. long. Staminate flowers ± crowded; perianth-lobes 2–3. Pistillate flowers less numerous; perianth tubular; stigmas 2. Endocarp-body tetrahedral, ± 2.5 mm. in diameter, tuberculate. Fig. 12.

TANZANIA. Morogoro District: Nguru Mts., NW. of Mkobwe, 29 Mar. 1953, *Drummond & Hemsley* 1901! & Uluguru Mts., Mwere valley, 26 Sept. 1970, *B.J. Harris et al.* 5125!; Songea District: Matagoro E. summit, 27 Mar. 1956, *Milne-Redhead & Taylor* 9352!
DISTR. T 6–8; Malawi
HAB. Rain-forest, often among rocks; 300–2000 m.

SYN. *D. hispida* Peter, F.D.O.-A. 2: 74, Descr. 4, t. 4 (1932), *non* Hook. (1840), *nom. illegit.* Type: Tanzania, Uluguru Mts., *Peter* 6893 (B, lecto.!)
D. kyimbilaensis De Wild., Pl. Bequaert. 6: 39 (1932); T.T.C.L.: 352 (1949). Type: Tanzania, Rungwe District, Mwakaleli, *Stolz* 2294 (BR, holo.!, K, UPS, Z, iso.!)

10. D. holstii *Engl.* in E.J. 20: 145 (1894) & E.M. 1: 13, t. 4A (1898); Rendle in F.T.A. 6(2): 36 (1916); Peter, F.D.O.-A. 2: 74 (1932); T.T.C.L.: 352 (1949). Type: Tanzania, Lushoto District, W. Usambara Mts., Mlalo, *Holst* 3766 (B, holo.!)

Herb (20–)30–90(–120) cm. tall, rhizomatous; stems erect, unbranched, up to 7 mm. thick, puberulous. Leaves in spirals; lamina chartaceous, oblong to elliptic, (4–)6–23(–29) × (2–)4–10 cm., apex acute to acuminate, base cuneate to subobtuse or sometimes

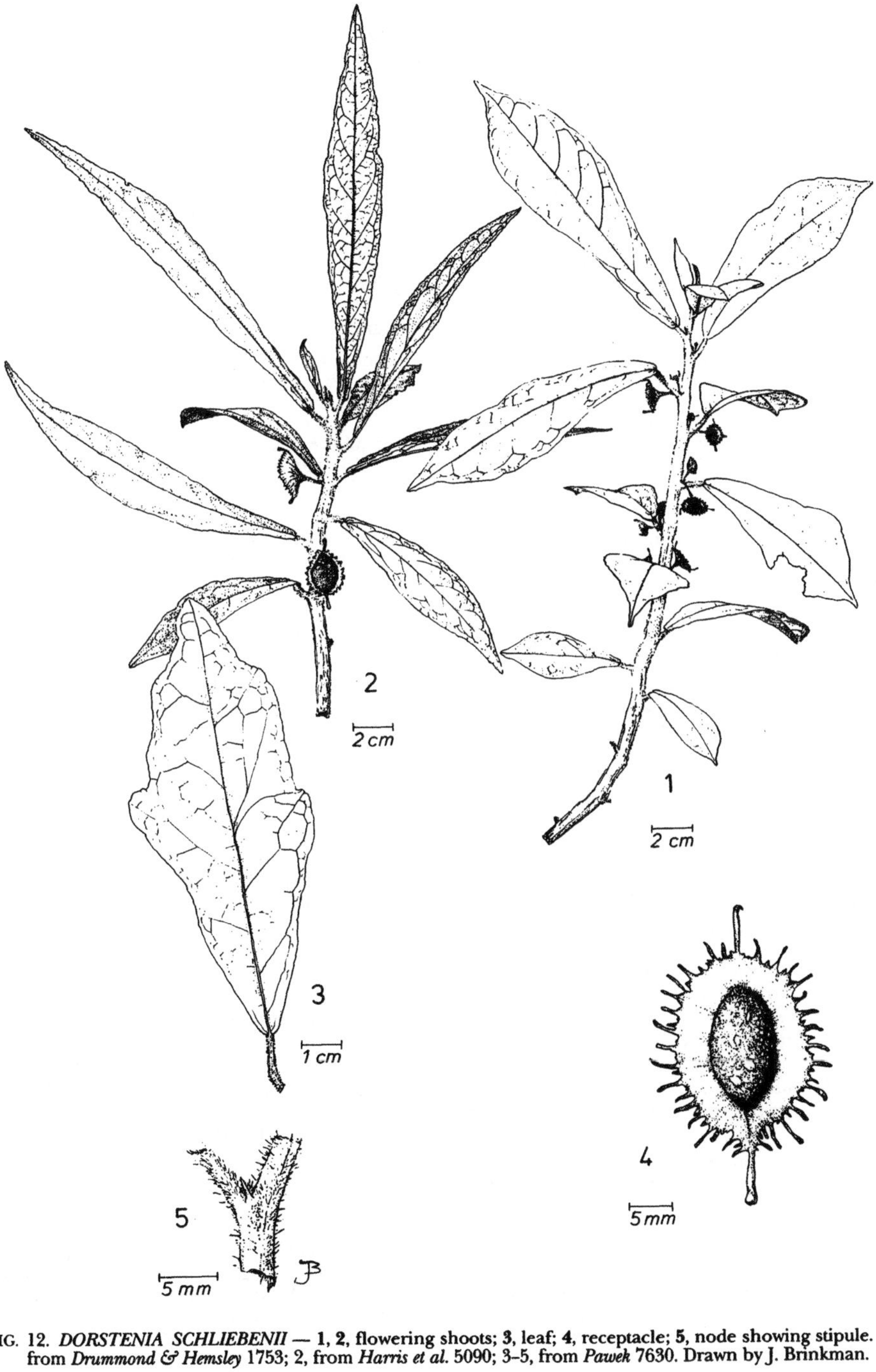

FIG. 12. *DORSTENIA SCHLIEBENII* — **1**, **2**, flowering shoots; **3**, leaf; **4**, receptacle; **5**, node showing stipule. 1, from *Drummond & Hemsley* 1753; 2, from *Harris et al.* 5090; 3–5, from *Pawek* 7630. Drawn by J. Brinkman.

subattenuate, margin entire, crenulate-dentate or sometimes deeply pinnately incised; upper surface glabrous, lower surface puberulous to sometimes hirtellous; lateral veins (4–)6–10 pairs; petiole 0.5–3.5 cm. long, 1–2 mm. thick; stipules subulate, up to 0.2 or 0.4–0.9 cm. long, subpersistent. Inflorescences solitary or in pairs; peduncle 0.5–4 cm. long, 1–1.5 mm. thick, broadened towards the receptacle. Receptacle ± funnel-shaped, orbicular, 0.7–2.5 cm. in diameter, margin 0.5–1 mm. wide; flowering face subcircular to multiangular or substellate, with 8–12(–14) 0.5–1.5 mm. long lobes; appendages in 1 row, up to 40, tooth-like, linear, spathulate or clavate, 1–40 mm. long. Staminate flowers crowded or spaced; perianth-lobes 3; stamens 3. Pistillate flowers among the staminate ones; perianth rather flat; stigmas 2. Endocarp-body ± 3 mm. in diameter, slightly tetrahedral, 2 sides tuberculate.

var. **holstii**

Stipules up to 2 mm. long. Peduncle 1–4 cm. long. Receptacle 0.7–2.5 cm. in diameter, substellate, with 8–12(–14) unequal triangular lobes passing into linear, up to 40 mm. long appendages, between the lobes 1 or 2 shorter appendages.

TANZANIA. Lushoto District: Amani, Jan. 1903, *Warnecke* 225! & 25 July 1953, *Drummond & Hemsley* 3442! & Derema [Nderema], 4 Apr. 1939, *Greenway* 5864!
DISTR. **T** 3; known only from the E. Usambara Mts.
HAB. Rain-forest, shady places, often among rocks; about 900 m.

SYN. *D. holstii* Engl. var. *grandifolia* Engl., E.M. 1: 13 (1898); Rendle in F.T.A. 6(2): 36 (1916); Peter, F.D.O.-A. 2: 75 (1932); T.T.C.L.: 352 (1949). Type: Tanzania, Lushoto District, E. Usambara Mts., Gonja [Gondja] Mt., *Buchwald* 236 (B, lecto.!, COI, K, isolecto.!)
D. usambarensis Engl. in E.J. 33: 114 (1902); Peter, F.D.O.-A. 2: 75 (1932). Type: Tanzania, Lushoto District, E. Usambara Mts., Ngwelo, Ngambo road, *Scheffler* 43 (B, holo.!)
D. holstii Engl. var. *lanceolata* Peter, F.D.O.-A. 2: 75 (1932), *nomen*, based on Tanzania, E. Usambara Mts., Kwamkoro–Amani, *Peter* 18693C (B!)

var. **longestipulata** *Hijman*, var. nov. stipulis longioribus, receptaculo suborbiculare vel multangulare appendicibus numerosis 1–13 mm. longis differt. Type: Tanzania, Morogoro District, Uluguru Mts., Diwale [Dualle] stream, *Schlieben* 4064 (B, holo.!, BR, K, LISC, MO, SRGH, iso.!)

Stipules 4–9 mm. long. Peduncle 0.5–1.2(–3) cm. long. Receptacle 0.7–1.9 cm. in diameter, subcircular to multiangular; appendages up to ± 40, close together, tooth-like, linear, spathulate or (especially the longer ones) clavate, 1–13 mm. long.

TANZANIA. Morogoro District: Nguru Mts., Manyangu Forest, Liwale valley, 27 Mar. 1953, *Drummond & Hemsley* 1849! & Nguru S. Forest Reserve, above Kwamanga, 5 Feb. 1971, *Mabberley & Pócs* 663; Pemba I., Chake–Weti road, 5 Aug. 1929, *Vaughan* 449!
DISTR. **T** 6; **P**; not known elsewhere
HAB. Forest, in wet shady places; near sea-level to 1500 m.

NOTE. The distinction between the two varieties is not always very marked, e.g. *Semsei* 1157 from the Nguru Mts., approximates towards var. *holstii*.

11. D. brownii *Rendle* in J.B. 53: 299 (1915) & in F.T.A. 6(2): 36 (1916); Hauman in F.C.B. 1: 62 (1948); U.K.W.F.: 318 (1974). Type: Uganda, Mengo District, Mabira Forest, *E. Brown* 460 (BM, holo.!)

Herb up to 50 cm. tall, rhizomatous; stems creeping, branched, 0.5–2 mm. thick, ± densely hirtellous. Leaves in spirals; lamina chartaceous when dry, elliptic to obovate, (1.5–)3–12 × (0.8–)2–5 cm., apex acute to faintly acuminate to subobtuse, base obtuse to subcordate, margin sinuate or towards the apex irregularly dentate; upper and lower surface hirtellous; lateral veins 4–7(–8) pairs; petiole 0.9–4 cm. long, ± 1 mm. thick; stipules subulate, 2–8 mm. long, persistent. Inflorescences 1–3-together; peduncle 0.7–2 cm. long, ± 0.5 mm. thick. Receptacle funnel-shaped; flowering face circular, 1–7 mm. in diameter, margin 0.1–0.5 mm. wide; appendages in 1 row, 12–18, triangular, 1–2.5 mm. long. Staminate flowers spaced; perianth-lobes (2–)3; stamens (2–)3. Pistillate flowers ± 8; perianth broadly conical; stigmas 2. Endocarp-body tetrahedral, 1–2 mm. in diameter, tuberculate.

UGANDA. Busoga District: Kagoma Forest Reserve, 5 Dec. 1952, *G.H. Wood* 535!; Mengo District: Mabira Forest, Mulange, Apr. 1919, *Dummer* 4055! & Kyagwe [Chagwe], Dec. 1922, *Maitland* 513!
KENYA. Kiambu District: Kamiti R. near junction with Ruiru–Kiambu road, 21 Dec. 1969, *Faden & Evans* 69/2073!; N. Kavirondo District: Kakamega Forest, near Forest Station, 12 Apr. 1973, *O.J. Hansen* 922!; Teita District: Mbololo Hill, Mraru Ridge, 12 Sept. 1970, *Faden et al.* 70/552!

TANZANIA. Morogoro District: Nguru Mts., Manyangu Forest, 2 Apr. 1953, *Drummond & Hemsley* 2003! & Mkobwe, 29 Mar. 1953, *Drummond & Hemsley* 1873! & NW. Uluguru Mts., 29 Sept. 1932, *Schlieben* 2752!
DISTR. **U** 3, 4; **K** 4, 5, 7; **T** 6; Zaire, Sudan, Ethiopia
HAB. Evergreen forest, wet places, often among rocks; 900–1700 m.

SYN. *D. penduliflora* Peter, F.D.O.-A. 2: 74, Descr. 4, t. 5/1a, b (1932). Type: Tanzania, Morogoro District, Uluguru Mts., Mkiri stream, *Peter* 32250 (B, holo.!)
D. ruwenzoriensis De Wild., Pl. Bequaert. 6: 59 (1932); Hauman in F.C.B. 1: 62 (1948). Type: Zaire, Ruwenzori Mts., Lanuri valley, *Bequaert* 4485 (BR, holo.!)

12. D. afromontana *R.E. Fries* in N.B.G.B. 8: 667 (1924); Friis in Norw. Journ. Bot. 21: 101–110 (1974); U.K.W.F.: 318 (1974). Type: Kenya, W. slopes of Mt. Kenya, *Fries* 1434 (UPS, holo.!, B, BM, BR, EA, MO, S, WAG, iso.!)

Herb 15–50 cm. tall, rhizomatous; stems erect, sometimes branched, 1–3 mm. thick, with short dense hairs in 2 longitudinal lines. Leaves in spirals; lamina papyraceous when dry, elliptic to obovate, sometimes slightly 3–5-lobed, (1–)2–9.5 × 1–4 cm., apex obtuse to acute, base cuneate to subobtuse, margin coarsely dentate-lobed; both surfaces hirtellous; lateral veins 2–6 pairs; petiole (0.5–)1–2.5 cm. long, ± 0.5 mm. thick; stipules subtriangular, 0.5–1 mm. long, subpersistent. Inflorescences solitary; peduncle 0.7–1 cm. long, 0.5–1 mm. thick, recurved, broadened towards the receptacle. Receptacle naviculate, in a vertical position, excentrically attached, the upper part 0.7–1 cm. long, the lower part 0.4–0.7 cm. long; flowering face 1–1.3 × 0.1 cm., margin 0.3 mm. wide; appendages in 1 row, primary ones 2, terminal, linear, the upper one 10–17 mm. long, the lower one 3–8 mm. long, secondary appendages lateral, numerous, tooth-like, 0.2–1 mm. long. Staminate flowers spaced; perianth-lobes 1 or 2; stamens 1 or 2. Pistillate flowers ± 7 in a median row; perianth tubular; stigmas 2. Endocarp-body ellipsoid, ± 2.5 × 2 mm., smooth or slightly tuberculate.

KENYA. Naivasha District: Aberdare Mts., 12 Mar. 1922, *Fries* 2183!; Kiambu District: Limuru Girls School, 22 Feb. 1970, *Faden & Evans* 70/74!; Meru District: NE. Mt. Kenya, Ithanguni, Kirui volcanic cone, 28 Feb. 1970, *Faden & Evans* 70/81!
DISTR. **K** 3, 4; not known elsewhere
HAB. Upland rain-forest; 2300–2600 m.

13. D. psilurus *Welw.* in Trans. Linn. Soc., Bot. 27: 71 (1869); Engl., E.M. 1: 20 (1898); Rendle in F.T.A. 6(2): 50 (1916); Hauman in F.C.B. 1: 69 (1948); Friis in Norw. Journ. Bot. 21: 101 (1974); Hijman in Fl. Cameroun 28: 74, t. 26 (1985). Type: Angola, Pungo Andongo, near Mata de Pungo, *Welwitsch* 1564 (BM, holo.!, B, G, K, P, iso.!)

Herb up to 60 cm. tall in Flora area but to 2(–3) m. in var. *scabra*, rhizomatous; rhizome often ± tuberous; stems erect or ascending, often branched, ± 3(–5) mm. thick, puberulous. Leaves in spirals, ± crowded at the apex of the stem; lamina papyraceous when dry, elliptic to obovate, (2–)5–19 × (1–)2–8 cm., apex acute to acuminate, base cuneate, margin denticulate to coarsely dentate or sometimes repand; both surfaces puberulous; lateral veins 5–8(–10) pairs; petiole 0.5–3.5 cm. long, 1–2 mm. thick; stipules narrowly triangular, 0.5–2(–4) mm. long, subpersistent. Inflorescences solitary or in pairs; peduncle 0.6–5 cm. long, ± 0.5 mm. thick. Receptacle vertical, naviculate; flowering face narrowly ovate to linear, 1.2–5.5 × 0.15–0.5 cm., margin almost lacking, sometimes up to 1.5 mm. wide; appendages 2, terminal, filiform, the upper one (20–)30–70(–100) mm. long, the lower one (1.2–)3–10(–30) mm. long. Staminate flowers ± spaced; perianth-lobes 1–2(–3); stamens 1–2(–3). Pistillate flowers 5–10(–22), most of them in a median row; perianth tubular; stigmas 2. Endocarp-body subglobose to tetrahedral, ± 3 mm. in diameter, at least slightly tuberculate.

UGANDA. W. Nile District: Kango, Apr. 1941, *Eggeling* 4266!; Bunyoro District: Budongo Forest, Bisu, May 1935, *Eggeling* 2023!; Mengo District: Mabira Forest, Nov. 1920, *Dummer* 4552!
TANZANIA. Buha District: Kakombe valley, 7 Jan. 1964, *Pirozyski* 184!; Kilosa, 30 Jan. 1926, *B.D. Burtt* 77!; Iringa District: Mufindi, Kigogo R., 19 Mar. 1962, *Polhill & Paulo* 1818!
DISTR. **U** 1, 2, 4; **T** 4, 6–8; Sudan west to Cameroun and south to Angola, Zambia, Zimbabwe and Mozambique
HAB. Forest and streamsides in woodland; 400–1350 m.

SYN. *D. bicornis* Schweinf. in Bot. Zeit. 29: 332 (1871), by mistake as *D. bicuspis* in Schweinf. in Bull. Mus. Hist. Nat. Paris, 1: 62 (1895); Rendle in F.T.A. 6(2): 49 (1916); Peter, F.D.O.-A. 2: 77 (1932). Type: Sudan, Niamniam, Tubami's Seriba, *Schweinfurth* 3788 (B, holo.!, BM, iso.!)

D. stolzii Engl. in E.J. 51: 432 (1914); Peter, F.D.O.-A. 2: 77 (1932). Type: Tanzania, Rungwe District, Kyimbila, *Stolz* 769 (B, holo.!, G, K, LD, M, S, iso.!)
D. psilurus Welw. var. *brevicaudata* Rendle in J.B. 53: 301 (1915) & in F.T.A. 6(2): 51 (1916). Type: Uganda, Bunyoro District, Budongo Forest, *Bagshawe* 931 (BM, holo.!)

NOTE. Specimens with relatively tall stems and relatively large leaves are mainly found in wet and shady places in rain-forest, whereas those with a thick tuberous rhizome and a denser indumentum on the stem and leaves are found in relatively dry habitats. Var. *scabra* Bureau, from forests of W. and central Africa, forms clumps with mostly unbranched stems up to 2(–3) m. tall.

14. D. goetzei *Engl.* in E.J. 28: 378 (1900); Rendle in F.T.A. 6(2): 47 (1916); Peter, F.D.O.-A. 2: 76 (1932). Type: Tanzania, Morogoro District, S. Uluguru Mts., *Goetze* 176 (B, holo.!)

Herb up to 40 cm. tall, rhizomatous; stems creeping, ascending or erect, unbranched or, especially creeping ones, branched, 1–5 mm. thick, ± fleshy, sparsely puberulous. Leaves in spirals; lamina chartaceous when dry, lanceolate to elliptic, 1–11 × 0.5–4.5 cm., apex acute to acuminate, base cuneate to subobtuse, margin repand to dentate; upper surface sparsely puberulous (more densely on the midrib), lower surface glabrous or sparsely puberulous (mainly on the midrib); lateral veins (2–)4–9 pairs; petiole (0.2–)0.7–3 cm. long, ± 0.5–2 mm. thick; stipules subulate, ± 0.5 mm. long, caducous. Inflorescences solitary; peduncle (0.5–)1–2.5 cm. long, ± 0.5 mm. thick. Receptacle discoid and plane; flowering face subcircular, 0.2–0.6 cm. in diameter, margin 1–3 mm. wide, membranous, with radiating ribs; appendages in 1 row, 1–4 triangular to linear, up to 10 mm. long. Staminate flowers spaced; perianth-lobes 2–3; stamens 2–3. Pistillate flowers numerous; perianth annular, flat; stigma 1. Endocarp-body tetrahedral, ± 1 mm. in diameter, tuberculate.

KENYA. Kilifi District: Cha Simba, 26 Aug. 1971, *Faden* 71/790! & 18 Aug. 1974, *B. Adams* 82!
TANZANIA. Lushoto District: Shume-Magamba Forest Reserve, Sungwi Peak, 11 July 1983, *Polhill et al.* 4982!; Morogoro District: Lukwangule Plateau, 17 Mar. 1953, *Drummond & Hemsley* 1634!; Ulanga District: Sali, Ngongo Mt., 22 Jan. 1979, *Cribb et al.* 11125!
DISTR. **K** 7; **T** 3, 6; not known elsewhere
HAB. Forest, in wet shaded places, often among rocks; 200–2000 m.

SYN. *D. goetzei* Engl. var *angustibracteata* Engl. in E.J. 28: 378 (1900); Rendle in F.T.A. 6(2): 47 (1916); Peter, F.D.O.-A. 2: 76 (1932). Type: Tanzania, S. Uluguru Mts., *Goetze* 182 (B, holo.)
D. liebuschiana Engl. in E.J. 46: 275 (1911); Rendle in F.T.A. 6(2): 73 (1916); Peter, F.D.O.-A. 2: 80 (1932). Type: Tanzania, Lushoto District, E. Usambara Mts., Lutindi, *Liebusch* (B, holo.!)

NOTE. The merging of the receptacle is diagnostic, but from the key characters above some specimens of *D. zanzibarica*, with the inner row of appendages obsolete, may be confused here.

15. D. tenuiradiata *Mildbr.* in N.B.G.B. 11: 1065 (1934). Type: Tanzania, Ulanga District, Mahenge, *Schlieben* 1700 (B, holo.!, G, P, S, Z, iso.!)

Herb up to 100 cm. tall, rhizomatous; stems ascending to erect, unbranched or branched, up to 7(–10) mm. thick, puberulous. Leaves in spirals; lamina chartaceous to papyraceous when dry, elliptic to subovate, (1–)3–13(–20) × (1–)2–5(–7) cm., apex acute to faintly acuminate, base cuneate to attenuate, margin irregularly, coarsely to faintly dentate; upper surface glabrous or sparsely puberulous on the midrib, lower surface glabrous or sparsely puberulous on the main veins; lateral veins 4–9 pairs; petiole 0.5–3(–4) cm. long, 0.5–1 mm. thick, mainly adaxially puberulous; stipules subulate, ± 1 mm. long, caducous. Inflorescences solitary; peduncle 1–2.5(–3) cm. long, ± 0.5 mm. thick. Receptacle discoid to slightly funnel-shaped, flowering face subcircular, 0.5–1.1 cm. in diameter, margin up to 0.5 mm. wide; appendages in 2 ± distinct rows, the inner (marginal) row triangular to linear, up to 8 mm. long, the outer (submarginal) row 8–12, filiform, up to 55(–70) mm. long. Staminate flowers ± spaced; perianth-lobes 3; stamens 3. Pistillate flowers numerous, spaced; perianth tubular; stigmas 2. Endocarp-body tetrahedral, ± 1 mm. in diameter, tuberculate.

TANZANIA. Morogoro District: Nguru Mts., 31 Jan. 1933, *Schlieben* 3357!; Ulanga District: Sali, Mbezi R., 24 Jan. 1979, *Cribb et al.* 11175!; Iringa District: Mufindi, Lulanda Forest Reserve, 17 Feb. 1979, *Cribb et al.* 11472!
DISTR. **T** 6, 7; not known elsewhere
HAB. On rocks in woodland or forest, often in wet places; 700–1625 m.

16. D. warneckei *Engl.* in E.J. 46: 275 (1911); Rendle in F.T.A. 6(2): 61 (1916); Peter F.D.O.-A. 2: 78 (1932). Type: Tanzania, Lushoto District, E. Usambara Mts., Sigi valley, *Warnecke* 509 (B, lecto.!, EA, iso.!)

Herb up to 100 cm. tall, rhizomatous; stems erect or ascending, branched (at least abortive branches present), 2–10 mm. thick, sparsely and minutely puberulous, the hairs mostly in longitudinal lines. Leaves in spirals; lamina papyraceous when dry, oblong to subobovate, less often lanceolate to elliptic, (1.5–)3–18(–24) × (0.3–)0.8–6 cm., apex acute to obtuse, base attenuate to cuneate, margin coarsely (dentate-)crenate to lobate or subentire; upper surface minutely puberulous on the midrib or also near the margin, lower surface glabrous or almost so; lateral veins 3–12 pairs; petiole 0.5–2.5 cm. long, ± 1–2 mm. thick; stipules triangular to ovate, up to 1 mm. long, persistent. Inflorescences solitary; peduncle 1–2.4 cm. long, ± 1 mm. thick. Receptacle discoid and face subcircular, 0.8–1.5 cm. in diameter; flowering face subcircular to slightly lobed, 0.7–1.4 cm. in diameter, margin up to 1 mm. wide; appendages basically in 2 rows, but at anthesis hardly distinguishable, the inner (marginal) row a few very short, the outer (submarginal) row ± 10–15, triangular to band-shaped or subspathulate, 0.5–10 mm. long. Staminate flowers ± spaced; perianth-lobes 3(–4); stamens 3–4. Pistillate flowers numerous; perianth shortly tubular; stigmas 2. Endocarp-body tetrahedral, ± 1.5 mm. in diameter, tuberculate.

KENYA. Kwale District: Shimba Hills, Sheldrick's Falls, 27 Dec. 1968, *Glover et al.* 1155!

TANZANIA. Tanga District: Magunga Estate, 28 Nov. 1952, *Faulkner* 1103!; Pangani District: Pangani R., between Hale and Makinyumbe, 1 July 1953, *Drummond & Hemsley* 3128!; Morogoro District: Uluguru Mts., above Morogoro, 16 Apr. 1935, *E.M. Bruce* 1059!

DISTR. **K** 7; **T** 3, 6; not known elsewhere

HAB. Forest, in shady and sometimes wet places, often among rocks; 200–750 m.

SYN. *D. latibracteata* Engl. in E.J. 46: 175 (1911). Type: Kenya, Kwale District, Cha Simba [Pemba], *Kassner* 388 (B, holo.!)

D. sacleuxii De Wild., Pl. Bequaert. 6: 61 (1932). Type: Tanzania, Tanga District, Amboni Forest, *Sacleux* 2319 (P, holo.!)

17. D. hildebrandtii *Engl.* in E.J. 20: 146 (1894) & E.M. 1: 23, t. 6b (1898); Rendle in F.T.A. 6(2): 65 (1916); Peter, F.D.O.-A. 2: 79 (1932). Type: Kenya, Teita District, Buchuma, *Hildebrandt* 2050 (B, holo.!)

Herb up to ± 70 cm. tall, rhizomatous (or tuberous); stems ascending to erect, branched (the branches often arrested, with minute leaves) or unbranched, fleshy to sometimes thickly succulent, the basal part often swollen, sometimes forming a globose to pear-shaped tuber up to 4 cm. across, the leafy part 1–5(–7) mm. in diameter, glabrous to sparsely puberulous; scars of the leaves, stipules and inflorescences often conspicuous and prominent. Leaves in spirals; lamina chartaceous to papyraceous when dry, oblong to elliptic, lanceolate, oblanceolate or linear, (0.5–)1–12(–20) × (0.3–)1–16 cm., apex obtuse to acute, base cuneate to attenuate or sometimes subobtuse; margin ± irregularly and sometimes coarsely crenate to dentate, sometimes subentire, when dry sometimes ± revolute; upper surface glabrous and smooth, scabridulous or sparsely puberulous on the midrib, lower surface puberulous to glabrous; lateral veins 2–10(–12) pairs; petiole (0.1–)0.2–2(–3) cm. long, 0.5–1 mm. thick, puberulous, mainly adaxially; stipules subulate to ovate, up to 1.5(–2) mm. long, persistent. Inflorescences solitary or sometimes in pairs; peduncle 0.2–1.2(–2.5) cm. long, ± 0.5 mm. thick, sometimes recurved. Receptacle discoid to broadly turbinate or funnel-shaped, sometimes subnaviculate,0.4–1.2 cm. in diameter (or wide); flowering face subcircular to substellate or narrowly elliptic to 3–6-angular, margin very narrow or lacking; appendages in 2 rows, the inner (marginal) row numerous, semi-circular to triangular, up to 2 mm. long, forming a crenate to dentate rim, outer (submarginal) row 2–12, narrowly triangular to band-shaped or filiform, (2–)5–22(–25) mm. long. Staminate flowers crowded or slightly spaced; perianth-lobes (2–)3(–4); stamens (2–)3(–4). Pistillate flowers numerous, scattered; perianth tubular; stigmas 2. Endocarp-body tetrahedral, 1–1.5 mm. in diameter, slightly tuberculate.

var. **hildebrandtii**

Herb up to 40 cm. tall; stems usually thickly succulent, at the base often swollen, tuber-like, mostly with several (normal and arrested) branches. Lamina usually chartaceous when dry, up to 6(–10) cm. long; lateral veins 3–6(–10) pairs; petiole up to 0.5(–1) cm. long. Peduncle 0.2–0.8(–1.2) cm. long. Flowering face subcircular, ± angular or elliptic; appendages of the inner row forming a dentate, crenate, subentire or obsolete rim; appendages of the outer (submarginal) row 2–8, up to 10 mm. long (shorter than the dimeter of the flowering face).

KENYA. Northern Frontier Province: Boni Forest, 2 Oct. 1947, *J. Adamson* 411 in *Bally* 5897!; Teita District: 17.6 km. Maungu Station to Rukanga road, 31 Dec. 1970, *Faden et al.* 70/955!; Kwale District: Mwachi, 10 Sept. 1953, *Drummond & Hemsley* 4254!

TANZANIA. Tanga District: Kange Forest, 27 Aug. 1955, *Faulkner* 1705!; Morogoro District: 14 km. NE. of Kingolwira Station, 7 Dec. 1957, *Welch* 452! & Kimboza Forest, 24 Jan. 1971, *Cribb & Grey-Wilson* 10414! & 1 Apr. 1954, *Padwa* 323!

DISTR. **K** 1; **T** 3, 6; not known elsewhere

HAB. Granitic, coral and limestone outcrops from open forest to woodland, bushland and succulent thickets, often near streams or in local water catchment areas, sometimes in shaded forest sites; 0–1000 m.

SYN. *D. braunii* Engl. in E.J. 46: 276 (1911); Rendle in F.T.A. 6(2): 66 (1916); Peter, F.D.O.-A. 2: 79 (1932). Type: Tanzania, Lushoto District, W. Usambara Mts., Shume–Mkumbara, *Braun* 2887 (B, holo.!)

D. tanneriana Peter, F.D.O.-A. 2: 78, Descr. 6, t. 6/1a, c (1932). Type: Tanzania, Tanga District, near Amboni, Ukereni Hill, *Peter* 25590 (B, lecto.!)

D. rhomboidea Peter, F.D.O.-A. 2: 79, Descr. 7, t. 8/2a, b & 9/2b (1932). Type: Tanzania, Handeni District, Kinamo–Sindeni, *Peter* 40650 (B, holo.!)

D. marambensis Peter, F.D.O.-A. 2: 78, Descr. 6, t. 7/2a, b (1932). Type: Tanzania, Lushoto District, E. Usambara Mts., Malamba [Maramba] to Kishangani, *Peter* 20761 (B, lecto.!)

D. brevifolia Peter, F.D.O.-A. 2: 74, Descr. 3, t. 3/2a–d (1932). Type: Tanzania, Lushoto District, W. Usambara Mts., near Mashewa, *Peter* 13244 (B, holo.!)

NOTE. On the basis of differences in the inflorescences two 'forms' can be distinguished.

a. Flowering face subcircular; appendages of the inner row forming a distinct dentate to crenate rim; appendages of the outer row 6–8, e.g. *Faulkner* 1705 & *Padwa* 323, cited above

b. Flowering face angular to elliptic; appendages of the inner row forming a faintly crenate to dentate, subentire or obsolete rim; appendages of the outer row 2–6, e.g. *Faden et al.* 70/955, *Drummond & Hemsley* 4254 & *Welch* 452, cited above.

var. **schlechteri** (*Engl.*) *Hijman*, stat. nov. Type: Mozambique, Beira, *Schlechter* (B, holo.!)

Herb up to 70 cm. tall; stems slender, at the base not swollen nor tuber-like, but slender and creeping, stems often unbranched. Lamina usually papyraceous when dry, up to 10(–19) cm. long; lateral veins 4–10(–12) pairs; petiole up to 2(–3) cm. long. Peduncle 0.5–1(–2) cm. long. Flowering face subcircular, ± quadrangular to substellate; appendages of the inner (marginal) row forming a distinct dentate to crenate rim, the appendages occasionally subulate and up to 2 mm. long. Fig. 13.

UGANDA. Bunyoro District: Budongo, Sonso R., Nov. 1935, *Eggeling* 2275!; Kigezi District: Mitano Gorge, Feb. 1947, *Purseglove* 2321!; Mengo District: Mabira Forest, Kiwala, June–July 1917, *Dummer* 3254!

KENYA. W. Suk District: Kapenguria, 15 May 1932, *Napier* 1915!; Fort Hall District: Chania gorge at Thika, 19 Sept. 1953, *Verdcourt* 1013!; Masai District: Talek R., 23 Aug. 1961, *Archer* 270!

TANZANIA. Bukoba District: Minziro Forest, Feb. 1959, *Procter* 1168!; Handeni District: Msangani R., 50 km. S. of Korogwe, 26 Dec. 1971, *Wingfield* 1799!; Morogoro District: Mgeta R., Hululu Falls, 15 Mar. 1953, *Drummond & Hemsley* 1578!

DISTR. **U** 2–4; **K** 1–7; **T** 1–7; Zaire, Rwanda, Burundi, Mozambique

HAB. On rocks, epiphyte on trees or with mosses and humus on wet ground in forests and along streams into woodland and thicket; 275–2100 m.

SYN. *D. schlechteri* Engl., E.M. 1: 23, t. 4e (1898); Rendle in F.T.A. 6(2): 61 (1916); Peter, F.D.O.-A. 2: 77 (1932); Hauman in F.C.B. 1: 71 (1948); U.K.W.F.: 319 (1974)

D. denticulata Peter, F.D.O.-A. 2: 74, Descr. 3, t. 3/1a, b (1932). Type: Tanzania, Lushoto District, W. Usambara Mts., near Mashewa, *Peter* 13793 (B, lecto.!)

D. polyactis Peter, F.D.O.-A. 2: 79, Descr. 8, t. 9/1a, b (1932). Type: Tanzania, Tabora District, Ngulu, SE. of Goweko, *Peter* 35111 (B, lecto.!)

D. carnulosa De Wild., Pl. Bequaert. 6: 27 (1932). Type: Tanzania, Tanga District, Amboni Forest, *Sacleux* 2542 (P, holo.!, BR, iso.!)

D. maoungensis De Wild., Pl. Bequaert. 6: 46 (1932). Type: Kenya, Teita District, Maungu [Maoungou] hill, *Sacleux* 2288 (P, holo.!)

D. renneyi Airy Shaw & Taylor in Cactus & Succ. Journ. Gr. Brit. 42: 87, t. (1980). Type: cultivated at Kew, No. 291-70-02846, from Kenya, Kitui District, Mutomo Hill, *Renney* (K, holo.!)

NOTE. This variety is altogether more mesic than var. *hildebrandtii* and tolerates less severe growing conditions, but is very widely distributed in eastern Africa.

On the basis of differences in the inflorescences three 'forms' can be distinguished.

a. Flowering face ± quadrangular to substellate; appendages of the outer row 4–8, up to 20 mm. long, at least longer than the diameter of the flowering face, e.g. *Eggeling* 2275, *Purseglove* 2321, *Procter* 1168 and *Drummond & Hemsley* 1578, cited above.

b. Flowering face subcircular; appendages of the outer row (6–)8–12, up to 20(–25) mm. long, at least longer than the diameter of the flowering face, e.g. *Dummer* 3254, *Napier* 1915 & *Archer* 270, cited above.

c. Flowering face subcircular; appendages of the outer row (6–)8–12, up to 10 mm. long, shorter than the diameter of the flowering face, e.g. *Verdcourt* 1013 & *Wingfield* 1799, cited above.

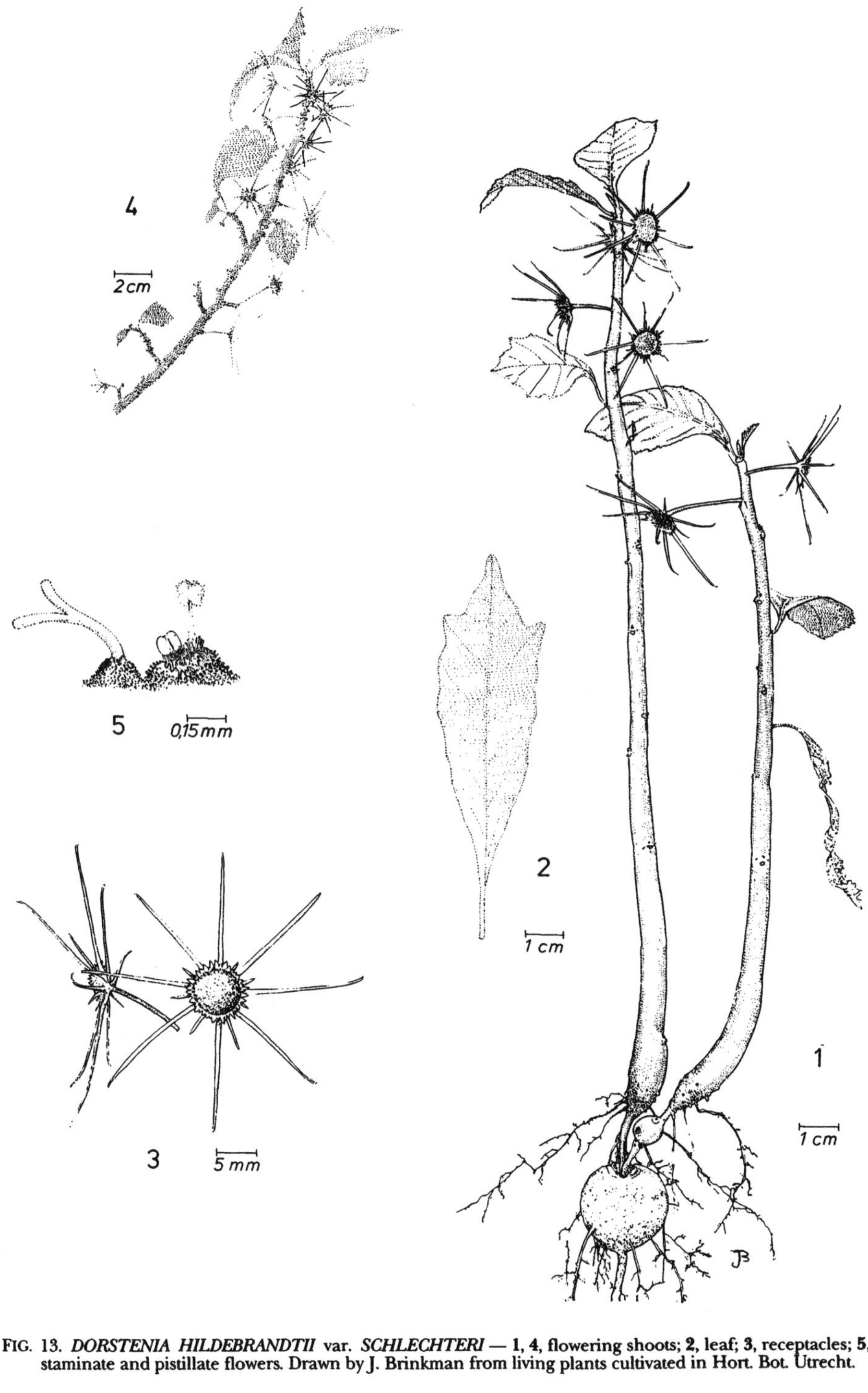

FIG. 13. *DORSTENIA HILDEBRANDTII* var. *SCHLECHTERI* — **1**, **4**, flowering shoots; **2**, leaf; **3**, receptacles; **5**, staminate and pistillate flowers. Drawn by J. Brinkman from living plants cultivated in Hort. Bot. Utrecht.

18. D. zanzibarica *Oliv.* in Hook., Ic. Pl. 16, t. 1581 (1887); Rendle in F.T.A. 6(2): 69 (1916); Peter, F.D.O.-A. 2: 80 (1932); U.K.W.F.: 319 (1974). Type: Tanzania, Zanzibar (see note), grown at Kew from material sent by Kirk (K, holo.!)

Herb up to 50(–75) cm. tall, rhizomatous or sometimes tuberous; stems erect or ascending, up to 100(–300) cm. long, the base sometimes forming a tuber up to 10 cm. across, branched (at least with arrested branchlets with minute leaves) or unbranched, 2–10 mm. thick when dry, sparsely minutely puberulous, the hairs mostly in ± distinct longitudinal lines. Leaves in spirals, often only at the top of the stem; lamina papyraceous when dry, oblong to subobovate to lanceolate or elliptic, (1–)2–14(–21) × (0.3–)1–4(–6.5) cm., apex acute to obtuse, base attenuate to cuneate, margin rather regularly, finely to coarsely crenate-dentate or sometimes repand; upper surface minutely (on the midrib rather densely) puberulous, lower surface puberulous on the veins or glabrous; lateral veins (5–)8–15 pairs; petiole (0.2–)1–2(–3) cm. long, 1–1.5 mm. thick; stipules triangular to ovate, up to 1 mm. long, persistent. Inflorescences often several together; peduncle (0.5–)1–3 cm. long, 0.5–1 mm. thick. Receptacle broadly turbinate to discoid, triangular to subquadrangular or sometimes ± irregularly substellate or oval, 0.5–1(–2) cm. in diameter, flowering face in outline similar to the receptacle, margin up to 1 mm. wide; appendages in 2 rows, the inner (marginal) row semi-circular, up to 1(–2) mm. long, forming a crenate rim, the outer (submarginal) row (2–)3–7, usually 3 or 3–4 of them band-shaped to triangular and 4–8 mm. long, the shorter ones (1–4) triangular to ovate, up to 3 mm. long. Staminate flowers ± crowded; perianth-lobes 3; stamens (2–)3. Pistillate flowers numerous; perianth shortly tubular; stigma 1 (sometimes a rudimentary second one). Endocarp-body tetrahedral, 1–±1.5 mm. in diameter, tuberculate. Fig. 14.

KENYA. Machakos/Masai District: Chyulu Hills, 25 Apr. 1938, *V.G. van Someren* 377 in *C.M.* 7686!; Teita District: Bura Bluff, 22 June 1964, *Archer* 493! & Maungu Hills, 31 May 1970, *Faden et al.* 70/152!

TANZANIA. Arusha District: Ngare Nanyuki, 31 Dec. 1970, *Richards & Arasululu* 26414!; Moshi District: Marangu, 12 Jan. 1945, *Bally* 4199!; Lushoto District: Mtai Escarpment, 25 May 1953, *Drummond & Hemsley* 2749!

DISTR. **K** 4/6, 7; **T** 2, 3, 6; ? **Z**; not known elsewhere

HAB. Rocks, cliffs, epiphytic or in ground humus, in various habitats from forest to succulent and deciduous bushland or thicket; up to 2400 m.

SYN. *D. volkensii* Engl. in E.J. 20: 143 (1894) & E.M. 1: 22, t. 7a (1898). Type: Tanzania, Moshi District, Marangu, *Volkens* 228 (B, lecto.!, BM, BR, K, isolecto.!)

D. holtziana Engl. in E.J. 51: 433 (1914); Rendle in F.T.A. 6(2): 69 (1916); Peter, F.D.O.-A. 2: 80 (1932). Type: Tanzania, Lushoto District, W. Usambara Mts., Mbalu-Kihurio[Kihuiro], *Engler* 1512 (B, lecto.!)

NOTE. The original material sent by Kirk from Zanzibar may well have been collected inland, as Bishop Hannington sent in material at the same time labelled Usagara.

19. D. buchananii *Engl.* in E.J. 20: 142 (1894) & E.M. 1: 23 (1898); Rendle in F.T.A. 6(2): 48 (1916); Peter, F.D.O.-A. 2: 76 (1932). Type: Malawi, without locality, *Buchanan* 505 (B, holo.!, K, iso.!)

Herb, tuberous; tuber discoid to subglobose, up to 4.5 cm. across; stems seasonal, erect, ascending or creeping, up to 50(–90) cm. long, 1–7(–10) mm. thick, puberulous, often branched (especially on creeping stems) or at least arrested branches transformed into subglobose tubers up to 1 cm. across. Leaves in spirals, on the lower part of the stem usually entire to tripartite scale leaves; lamina papyraceous when dry, oblong to elliptic or subobovate, sometimes subcircular, (0.5–)2–15 × (0.3–)1–5 cm., apex obtuse to acute or subacuminate, sometimes rounded, base cuneate to attenuate or obtuse, margin ± irregularly rather coarsely dentate-crenate to denticulate; upper surface smooth, puberulous to strigillose, lower surface puberulous; lateral veins 4–12(–14) pairs; petiole 0.3–2 cm. long, 1–2 mm. thick; stipules narrowly triangular to subulate, up to 1.5(–4) mm. long, persistent. Inflorescences solitary; peduncle 0.2–1(–1.6) cm. or 4–9 cm. long. Receptacle broadly turbinate and in outline elliptic or subnaviculate, 1–2.3 × 0.5–1(–1.2) cm.; flowering face narrowly elliptic, margin very narrow or lacking; appendages in 2 rows, the inner (marginal) row semi-circular, up to 0.5(–1) mm. long, the outer (submarginal) row 2, terminal, band-shaped to filiform, 15–100(–130) mm. long. Staminate flowers ± crowded; perianth-lobes 2; stamens 2. Pistillate flowers numerous; perianth tubular; stigmas 2. Endocarp-body tetrahedral, ± 1.5 mm. in diameter, tuberculate.

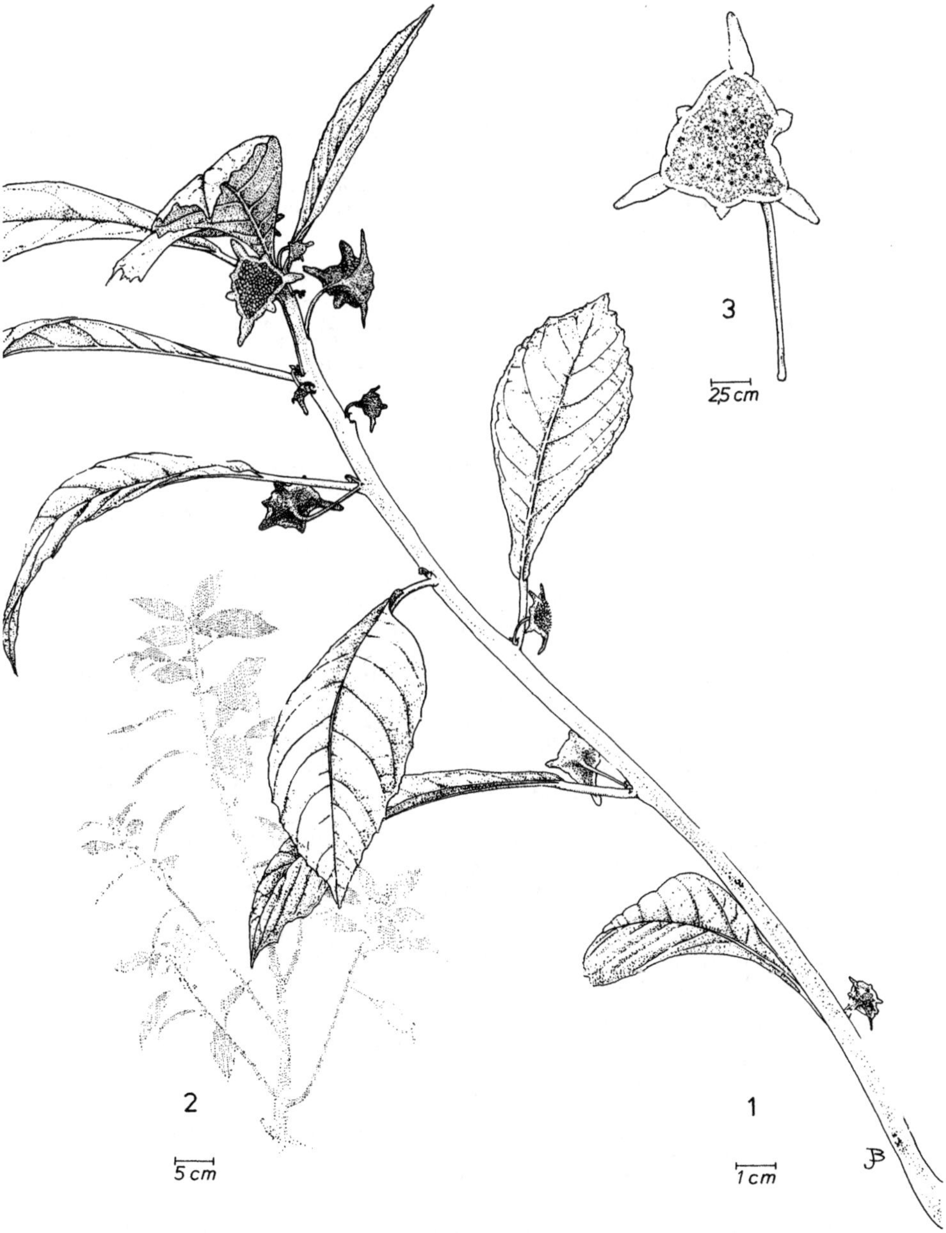

FIG. 14. *DORSTENIA ZANZIBARICA* — **1**, flowering shoot; **2**, habit; **3**, receptacle. Drawn by J. Brinkman.

var. **buchananii**

Stems erect or creeping. Lamina elliptic, 2–6 × 1.5–3.5 cm. Peduncle ± 1 cm. long. Receptacle ± 1.5–2 × 1 cm.; upper appendage ± 80 mm. long, lower one ± 15 mm. long.

TANZANIA. Chunya District: S. Rukwa Escarpment above Bangala [Mbangala], 14 Dec. 1963, *Richards* 18666!
DISTR. T 7; Malawi, Zambia, Zimbabwe and Mozambique
HAB. Rock outcrops and streamsides in *Brachystegia* woodland and riverine communities; 900 m.

SYN. *D. caudata* Engl. in E.J. 28: 377 (1900); Rendle in F.T.A. 6(2): 67 (1916); Peter, F.D.O.-A. 2: 79 (1932). Type: Tanzania, Iringa District, Ruaha R., *Goetze* 459 (B, holo.!)
D. unicaudata Engl. in E.J. 51: 432 (1914). Type: Tanzania, E. Usambara Mts., Kifalufalu, *Herb. Amani* 2544 (B, holo., presumably destroyed)

var. **longepedunculata** *Rendle* in J.B. 53: 302 (1915) & in F.T.A. 6(2): 48 (1916). Type: Mozambique, near Lake Malawi [Nyassa], *Johnson* 494 (K, holo.!)

Stems erect. Lamina elliptic to oblong, (5–)6–15 × (1–)2–5 cm. Peduncle 4–9 cm. long. Receptacle ± 1–1.3 × 0.5–1 cm.; upper appendage 80–130 mm. long, lower appendage 15–70 mm. long.

TANZANIA. Mpwapwa, 22 Jan. 1935, *Hornby* 614!; Iringa District: Madibira village, 12 Dec. 1961, *Richards* 15596!; Tunduru District: 11 km. E. of Songea District boundary, near Libobi village, 21 Dec. 1955, *Milne-Redhead & Taylor* 7739!
DISTR. T 5, 7, 8; Malawi, Zambia and Mozambique
HAB. As var. *buchananii*; 200–1700 m.

SYN. *D. ruahensis* Engl. in E.J. 28: 377 (1900); Rendle in F.T.A. 6(2): 67 (1916); Peter, F.D.O.-A. 2: 79 (1932). Type: Tanzania, Iringa District, Ruaha R., *Goetze* 429 (B, holo.!)
D. ruahensis Engl. var. *appendiculosa* Peter, F.D.O.-A. 2: 79 (1932), *nom. subnud.*, based on Tanzania, Dodoma District, Itigi–Bangayega, *Peter* 33847 (B†)
D. longipedunculata De Wild., Pl. Bequaert. 6: 44 (1932); T.T.C.L.: 352 (1949). Type: Tanzania, Dodoma District, Sagara, *Sacleux* 2234 (P, holo.!)

20. D. benguellensis *Welw.* in Trans. Linn. Soc., Bot. 27: 71 (1869); Rendle in F.T.A. 6(2): 64 (1916); Peter, F.D.O.-A. 2: 79 (1932); Hauman in F.C.B. 1: 74 (1948); Hijman in Fl. Cameroun 28: 94, t. 32/5–7 (1985). Type: Angola, Huila, *Welwitsch* 1566 (K, holo., G, iso.!)

Herb up to 50(–60) cm. tall, tuberous; tuber discoid to subglobose, 1–12 cm. in diameter; stems annual, unbranched or branched (mostly with arrested branches), 1.5–8 mm. thick, puberulous to hirtellous or hispidulous. Leaves in spirals, on the lower part of the stem with entire to tripartite scale leaves; lamina of foliage leaves chartaceous when dry, oblong to subovate or linear, sometimes subobovate, occasionally elliptic or ovate, 1–15 × 0.2–4.5 cm., apex acute to subacuminate, base obtuse to rounded, margin finely to coarsely dentate or occasionally subcrenate; both surfaces puberulous to hirtellous or hispidulous; lateral veins 4–12 (in linear leaves up to 25) pairs; petiole (0–)0.1–0.2(–0.5) cm. long, ± 1 mm. thick; stipules triangular to oblong, up to 5 mm. long, persistent. Inflorescences solitary or occasionally in pairs; peduncle 0.5–7 cm. long, 1–1.5 mm. thick. Receptacle discoid to broadly turbinate to shallowly cup-shaped, subcircular, 0.5–2(–2.5) cm. in diameter; margin very narrow; appendages in 2 rows, inner (marginal) row with numerous triangular to subulate or ovate appendages up to 1–2 mm. long, or the appendages indistinct and the rim subentire, outer (submarginal) row mostly with 5–12 (less commonly with up to 23 or with less than 5, down to 2, 1 or even 0), band-shaped to filiform, less often subspathulate to oblong, (1–)2–20(–80) mm. long appendages. Staminate flowers ± crowded; perianth-lobes 2; stamens 2. Pistillate flowers numerous; perianth shortly tubular; stigmas (1–)2. Endocarp-body tetrahedral, ± 2 mm. in diameter, tuberculate, pale brown. Fig. 15/5–7.

UGANDA. Karamoja District: Napak, June 1950, *Eggeling* 5924!; Teso District: Serere, May 1932, *Chandler* 608!; Mbale District: Buligenyi [Bulingenyi], 19 May 1955, *Norman* 267!
KENYA. Elgon, *Jack* 262!; Uasin Gishu District: Kipkarren, *Brodhurst-Hill* 146!
TANZANIA. Bukoba District: Ndama, Oct. 1931, *Haarer* 2313!; Ufipa District: Chapota, 5 Dec. 1949, *Bullock* 2060!; Songea District: 16 km. W. of Songea, 2 Jan. 1956, *Milne-Redhead & Taylor* 8106!
DISTR. U 1, 3; K 3; T 1, 4, 5, 7, 8; Sudan and westwards to Cameroun, southwards through Zaire to Angola, Zambia, Zimbabwe, Malawi and Mozambique
HAB. Woodland, wooded and open grassland, often in local water catchment areas in valleys or among rocks; 1000–2450 m.

SYN. *D. poggei* Engl. in E.J. 20: 146 (1894). Type: Angola, R. Cuango [Quango], *Pogge* 294 (B, holo.!)
D. debeerstii De Wild. & Th. Dur. in B.S.B.B. 39: 75 (1900); Rendle in F.T.A. 6(2): 62 (1916); Peter, F.D.O.-A. 2: 78 (1932). Type: Zaire, Shaba, Marungu Mts., *Debeerst* (BR, holo.!)

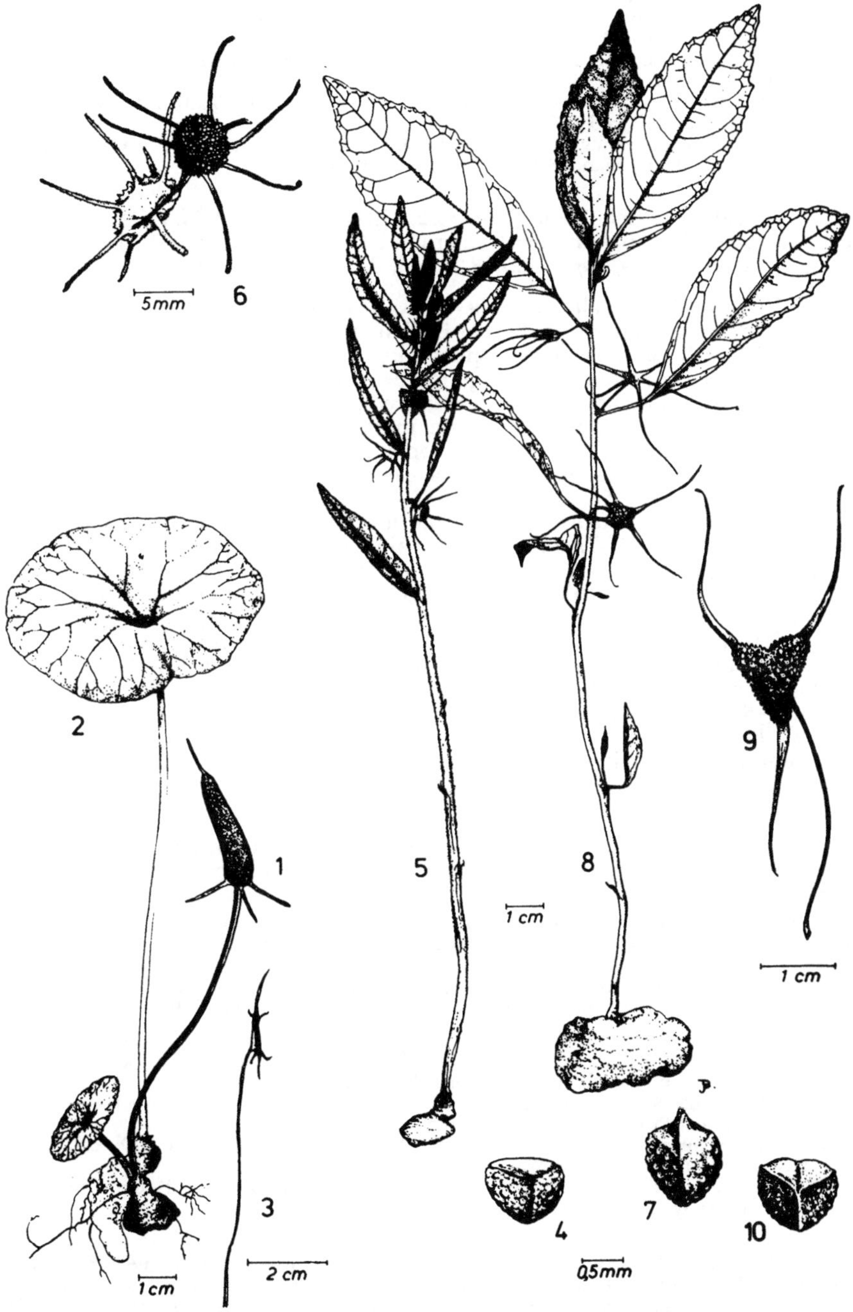

FIG. 15. *DORSTENIA BARNIMIANA* VAR. *TROPAEOLIFOLIA* — **1, 2,** habit; **3,** inflorescence; **4,** endocarp body. *D. BENGUELLENSIS*—**5,** habit; **6,** inflorescences; **7,** endocarp body. *D. CUSPIDATA*—**8,** habit; **9,** infructescence; **10,** endocarp-body. 1, from *Mildbraed* 9433; 2, 3, from *Ledermann* 3854; 4–6, from *Mildbraed* 9382; 7, from *Jacques-Félix* 3944; 8, from *Le Testu* 3055; 9, 10, from *Keay* in *FHI* 22928. Drawn by J. Brinkman.

D. rosenii R.E. Fries in Arkiv Bot. 13(1): 15 (1913). Type: Zambia, Mukanshi R., *R.E. Fries* 1125 (UPS, holo.!)
D. rosenii R.E. Fries var. *multibracteata* R.E. Fries in Arkiv Bot. 13(1): 15 (1915). Type: Zambia, Mbala [Abercorn], *R.E. Fries* 11259 (UPS, holo.!)
D. poggei Engl. var. *meyeri-johannis* Engl. in E.J. 51: 434 (1914); Peter, F.D.O.-A. 2: 79 (1932), under *D. benguellensis*. Type: Tanzania, Kahama District, Ushirombo [Uschirombo], *Meyer* (B, holo., presumably destroyed)
D. debeerstii De Wild. & Th. Dur. var. *multibracteata* (R.E. Fries) Rendle in F.T.A. 6(2): 63 (1916); Peter, F.D.O.-A. 2: 79 (1932)
D. mildbraediana Peter, F.D.O.-A. 2: 79, Descr. 7, t. 8/1 (1932). Type: Tanzania, Dodoma District, Chaya [Tschaya], *Peter* 34080 (B, lecto.!)
D. achtenii De Wild., Contr. Fl. Katanga, Suppl. 4: 100 (1932); Hauman in F.C.B. 1: 75 (1948). Type: Zaire, Shaba, Biano, *Achten* 1 (BR, holo.!, B, iso.!)

21. D. cuspidata A. Rich., Tent. Fl. Abyss. 2: 272 (1850); Engl., E.M. 1: (1898); Rendle in F.T.A. 6(2): 60 (1916); Hijman in Fl. Cameroun 28: 92, t. 32/9, 10 (1985). Type: Ethiopia, Tigray, Djeladjeranne, *Schimper* 1727 (P, holo.!, B, K, L, S, UPS, iso.!)

Herb 10–40(–50) cm. tall, tuberous or with swollen rhizomes; tuber discoid to subglobose or irregular in shape, 1–4 cm. in diameter; stems annual, mostly unbranched 1–4 mm. thick, puberulous. Leaves in spirals, on the lower part of the stem with entire or tripartite scale leaves; lamina of foliage leaves papyraceous to chartaceous when dry, elliptic to oblong or subobovate, less often lanceolate or linear, 1.5–12 × 0.5–5.5 cm., apex obtuse to acute, base cuneate to attenuate, margin finely crenate to dentate, coarsely dentate-crenate or repand to subentire; upper surface puberulous, at least in part, to substrigillose or subhispidulous, lower surface puberulous to hirtellous; lateral veins 3–12 (or in linear leaves up to 25) pairs; petiole (0.2–)0.5–2.5 cm. long, 1–1.5 mm. thick; stipules narrowly triangular to subulate, (0.5–)2–5 mm. long, persistent. Inflorescences solitary or in pairs; peduncle 0.8–3(–7) cm. long, 0.5–1 mm. thick. Receptacle discoid to broadly turbinate, ± irregularly stellate to subquadrangular, triangular or subcircular, 0.7–2.2 cm. in diameter or (if subnaviculate) 1–1.5 × 0.3–0.5 cm.; flowering face ± similar to the receptacle in outline, margin very narrow to absent; appendages in 2 rows, the inner (marginal) row numerous, triangular to ovate, up to 1 mm. long, or indistinct and the rim subentire, outer (submarginal) row 2 (in subnaviculate receptacles), 3 (in triangular receptacles) or mostly 4–14, narrowly triangular to band-shaped or less often filiform, (5–)20–60(–85) mm. long. Staminate flowers crowded to ± spaced; perianth-lobes 2; stamens 2. Pistillate flowers numerous; perianth shortly tubular; stigma 1. Endocarp-body tetrahedral, 1.2–1.5 mm. in diameter, tuberculate, greyish. Fig. 15/8–10, p. 39.

var. **cuspidata**

Stems erect from globose to discoid tubers 1–7 cm. across. Leaves with 8–12(–25 in linear leaves) pairs of lateral veins. Receptacle broadly turbinate, stellately lobed with (3–)4–8(–14) lobes, and (3–)4–8(–14) appendages.

TANZANIA. Mpanda District: Kungwe Mt., Dec. 1956, *Procter* 642!; Ufipa District: Kala Bay, 31 Dec. 1963, *Richards* 18724!; Tunduru District: 11 km. E. of Songea District boundary, 21 Dec. 1955, *Milne-Redhead & Taylor* 7855!
DISTR. **T** 4, 5, 7, 8; Senegal to Ethiopia and south to Zambia, Zimbabwe, Mozambique and Malawi
HAB. Deciduous woodland, often among rocks or in shady places; 750–1200 m.

SYN. *D. walleri* Hemsley in Gard. Chron., ser. 3, 14: 178 (1893); Rendle in F.T.A. 6(2): 68 (1916); Peter, F.D.O.-A. 2: 80 (1932); F.W.T.A., ed. 2, 1: 599 (1958). Type: cultivated at Kew from material from Malawi, Shire Highlands, Nakatupa, *Buchanan* (K, holo.!)
D. unyikae Engl. & Warb. in E.J. 30: 291 (1901). Type: Tanzania, Mbeya District, Unyiha, [Unyika], near Piseki, *Goetze* 1422 (B, holo.!)
D. walleri Hemsley var. *minor* Rendle in F.T.A. 6(2): 68 (1916); Peter, F.D.O.-A. 2: 80 (1932). Based on *D. unyikae* Engl. & Warb.
D. tetractis Peter, F.D.O.-A. 2: 80, Descr. 8, t. 10/1 (1932). Type: Tanzania, Dodoma District, Lake Chaya [Tschaya], *Peter* 34383 (B, holo.!)

var. **brinkmaniana** *Hijman*, var. nov. a var. *cuspidata* nervis foliorum utrinque 5–8, receptaculo disciforme, subtriangulare ad angulare vel suborbiculare differt. Type: Tanzania, Iringa District, Ruaha National Park, Trekimboga [Treakimboga] track, *Greenway & Kanuri* 14786 (EA, holo.!, K, M, MO, iso.!)

Succulent plants up to 35 cm. tall, with erect stems arising from globose to discoid tubers 1.5–3 cm. across. Leaves with 5–8 pairs of lateral veins. Receptacle discoid, subtriangular to angular or subcircular with 3–10 appendages.

KENYA. Machakos District: Koboko, 5 km. S. of Hunter's Lodge, 26 Dec. 1960, *Archer* 202!; Kilifi, 16 Aug. 1973, *Mason* in *E.A.H.* 15404!
TANZANIA. Mwanza District: Beda, 15 Mar. 1933, *Haarer* 306!; Mpwapwa District: Kikombo R., 22 Apr. 1932, *B.D. Burtt* 3856!; Mbeya District: Utengule–Igurusi, Jan. 1963, *Procter* 2275!
DISTR. **K** 4, 7; **T** 1, 5–7; N. Zambia
HAB. Damp shaded places in deciduous and coastal bushland, thickets, wooded grassland and along rivers; near sea-level to 1200 m.
DISTR. (of species as a whole). **K** 4, 7; **T** 1, 4–8; Senegal to Ethiopia and south to Zambia, Zimbabwe and Mozambique, also in Madagascar

NOTE. A polymorphic species subdivided with a number of varieties to be published shortly in K.B. In East Africa var. *cuspidata* belongs to the form formerly known as *D. walleri* and characteristic of the *Brachystegia* woodlands of Tanzania and southwards, whereas var. *brinkmaniana* is characteristic of the thickets, bushland and wooded grasslands of the Somali–Masai region.

22. D. foetida (*Forssk.*) *Schweinf.* in Bull. Herb. Boiss. 4, App. 2: 120 (1896); Engl., E.M. 1: 26 (1898); Rendle in F.T.A. 6(2): 72 (1916); U.K.W.F.: 320 (1974); Friis in Nordic Journ. Bot. 3: 536, fig. 3 (1983). Type: N. Yemen, Hadie, *Forsskål* (C, holo., K, photo.!)

Undershrubs up to 15 cm. tall; stems succulent, erect, unbranched or branched, up to 8 mm. thick, puberulous or glabrous, internodes short, periderm of older parts flaking off, scars of leaves, stipules and inflorescences conspicuous and prominent. Leaves in spirals, subrosulate; lamina chartaceous to papyraceous when dry, oblong to narrowly lanceolate or elliptic, 3.5–14 × 0.3–2 cm., apex acute, base attenuate, margin ± undulate and crenate to dentate or subentire; both surfaces minutely puberulous; lateral veins 5–15 pairs; petiole 0.1–2.5 cm. long, ± 1 mm. thick; stipules subulate, 1–4 mm. long, subpersistent or caducous. Inflorescences solitary; peduncle 1–3 cm. long, ± 1 mm. thick. Receptacle discoid, 0.5–1 cm. in diameter; flowering face substellate, margin almost lacking; appendages in 2 rows, the inner (marginal) row several, tooth-like, up to 1 mm. long, the outer (submarginal) row 5–10, band shaped to filiform, 2–12 mm. long. Staminate flowers ± spaced; perianth-lobes 2; stamens 2. Pistillate flowers numerous; perianth shortly tubular; stigma 1. Endocarp-body tetrahedral, ± 1.5 mm. in diameter, tuberculate.

KENYA. Northern Frontier Province: Dandu, 27 Mar. 1952, *Gillett* 12630!; Machakos District: Kyulu Station, 10 Jan. 1964, *Verdcourt* 3880!; Tana River District: Garissa–Garsen, 7–9 July 1974, *Faden* 74/1017!
TANZANIA. Pare District: Pangani R. at Nyumba ya Mungu, Oct. 1964, *Beesley* 51!
DISTR. **K** 1, 4, 7; **T** 3; Ethiopia, Somalia, Saudi Arabia, Yemen and Oman
HAB. Rock outcrops and open places in deciduous and succulent bushland; 100–1000 m.

SYN. *Kosaria foetida* Forssk., Fl. Aegypt.-Arab.: 164 (1775)
Cosaria forskahlii J. Gmelin in Syst. Nat., ed. 13, 2 (1): 71 (1796)
Dorstenia obovata A. Rich., Tent. Fl. Abyss. 2: 272 (1850). Type: Ethiopia, Tacazze, Djeladjeranne, *Schimper* 1675 (B, lecto.!, M, MO, isolecto.!)
D. crispa Engl., E.M. 1: 27, t. 9A (1898); Rendle in F.T.A. 6(2): 73 (1916). Type: Ethiopia, Sidamo, Gerima on R. Daua, *Riva* 441 (B, holo.!, FT, iso.)
D. foetida (Forssk.) Schweinf. var. *obovata* (A. Rich.) Engl., E.M. 1: 27, 1G (1898)
D. crispa Engl. var. *lancifolia* Rendle in J.B. 53: 302 (1915) & in F.T.A. 6(2): 73 (1916). Lectotype — see Friis (1983): Kenya, Machakos District, Ngomeni, *Scott Elliot* 6279 (BM, lecto.!, K, isolecto.!)
D. foetida (Forssk.) Schweinf. subsp. *lancifolia* (Rendle) Friis in Nordic Journ. Bot. 3: 538 (1983)

NOTE. *D. foetida* is very variable in habit, in the shape and size of the leaves, length of the petioles and stipules. Rather tall unbranched stems often bear lanceolate leaves. Rather short, branched stems often bear broad, obovate to subcircular leaves, but intermediates between these types occur. The lanceolate-leaved form, recognised as subsp. *lancifolia* by Friis (1983), is the most usual form in Kenya except along the Ethiopian border and sporadically elsewhere, e.g. *Verdcourt* 3880, cited above.

23. D. barnimiana *Schweinf.*, Pl. Nilot.: 36, t. 12 (1862); Engl., E.M. 1: 24 (1898); Rendle in F.T.A. 6(2): 70 (1916); Hauman in F.C.B. 1: 80 (1948); U.K.W.F.: 320, fig. on p. 319 (1974); Friis in Nordic Journ. Bot. 3: 534, fig. 1 (1983); Hijman in Fl. Cameroun 28: 96, fig. 32/1–4 (1985). Type: Sudan, Sennar, Fazogli [Fazugli], *Hartmann* (B, holo.†)

Herb up to ± 25 cm. tall, tuberous (or ± rhizomatous); tuber subglobose to pyriform, up to 4 cm. in diameter, gradually passing into a stem up to 1.5(–2.5) cm. long and 1 cm. thick with short internodes. Leaves subrosulate; lamina papyraceous to chartaceous when dry, peltate or basally attached, circular to broadly ovate or cordate, entire or palmately incised down to the petiole (with segments up to 12 × 2 cm.), 2.5–12 × 3.5–10 cm., apex rounded to subacute, base cordate, margin entire, crenate, denticulate or sinuate; upper surface puberulous, mainly towards the margin, lower surface puberulous; in basally attached leaves lateral veins 2–3 pairs, in peltate leaves 5–6 radiating veins; petiole up to 15 cm. long, 1–5(–8) mm. thick; stipules subulate, up to 1 mm. long, caducous. Inflorescences solitary; peduncle (3–)4.5–16(–26) cm. long, 1–3 mm. thick. Receptacle in a vertical position, naviculate to subligulate, 1.2–4.5 × 0.2–1 cm.; flowering face linear to subovate or elliptic, margin up to 1 mm. wide; appendages in 1 row, linear to filiform, 1 apical appendage, up to 80 mm. long, lateral appendages 2 (near base) or several to numerous, up to 4.5 cm. long. Staminate flowers crowded; perianth-lobes 2; stamens 2. Pistillate flowers numerous; perianth cushion-shaped; stigma 1 (sometimes a second rudimentary stigma). Endocarp-body tetrahedral, ± 1.5 mm. in diameter, slightly tuberculate.

var. **barnimiana**

Lamina basally attached, subcircular, cordate to broadly ovate, entire or palmately incised, up to 12 × 10 cm.; petiole (3–)6–15 cm. long, 1–3 mm. thick. Peduncle 0.8–26 cm. long, 1–2 mm. thick; lateral appendages 2–±50.

UGANDA. Acholi District: Imatong Mts., Apr. 1938, *Eggeling* 3602!; Karamoja District: Akisim–Napak, June 1950, *Eggeling* 5883!; Mbale District: Sebei, 7 June 1953, *Norman* 224!
KENYA. Northern Frontier Province: Moyale, 16 Apr. 1952, *Gillett* 12798!; Machakos District: N. side of Thika R., S. of Mabaloni Hill, 23 May 1970, *Faden & Evans* 70/128!; Kisumu-Londiani District: Lumbwa, 21 Apr. 1922, *Fries* 2796!
TANZANIA. Moshi District: W. Kilimanjaro, Rongai Ranches, 19 Apr. 1957, *Greenway* 9184!; Lushoto District: W. Usambara Mts., Shume Forest, 3 km. SE. of Manolo, 23 May 1953, *Drummond & Hemsley* 2694!; Mpanda District: Kapapa–Uruwira, 30 Oct. 1959, *Richards* 11666!
DISTR. **U** 1–4; **K** 1, 3–6; **T** 2–4, 6, 7; Cameroun to Somalia and Yemen (N. & S.), also in Zambia
HAB. Wooded and open grassland, often on shallow soils overlying rock; (600–)1000–2300 m.

SYN. *D. ophioglossoides* Bureau in DC., Prodr. 17: 276 (1873). Type: Ethiopia, Semien, Debra Eski, *Schimper* 402 (B, holo.!, FT, G, K, S, iso.!)
D. telekii Schweinf. in Engl., Hochgebirgsfl. Trop. Afr.: 190 (1892). Type: Kenya, Kikuyu, *von Höhnel* 28 (B, holo.!)
D. palmata Engl. in E.J. 20: 146 (1894). Type: Sudan, Seriba Ghattas, *Schweinfurth* 1881 (B, holo.!, G, K, iso.!)
D. barnimiana Schweinf. var *ophioglossoides* (Bureau) Engl., E.M. 1: 25 (1898); Rendle in F.T.A. 6(2): 71 (1916)
D. barnimiana Schweinf. var *telekii* (Schweinf.) Engl., E.M. 1: 25 (1898); Rendle in F.T.A. 6(2): 71 (1916)
D. barnimiana Schweinf. var. *angustior* Engl. in E.J. 30: 292 (1902); Rendle in F.T.A. 6(2): 71 (1916); Peter, F.D.O.-A. 2: 80 (1932). Type: Tanzania, Mbeya District, Unyiha [Unyika], near Toola, *Goetze* 1415 (B, holo.!, BM, iso.!)

NOTE. Flowers appear after the beginning of the rainy season; a short time later the leaves appear; young leaves are often shortly petiolate and appressed to the ground.

var. **tropaeolifolia** (*Schweinf.*) *Rendle* in F.T.A. 6(2): 71 (1916); Hijman in Fl. Cameroun 28: 95, t. 32/1–4 (1985). Type: Ethiopia, Gendoa R., *Schweinfurth* 564 (B, holo., G, P, U, iso.!)

Lamina peltate, subcircular, 3–7.5 cm. in diameter; petiole 2–8(–14) cm. long, 1–5(–8) mm. thick. Peduncle 8–13.5 cm. long, 1–4 mm. thick; lateral appendages 2–14. Fig. 15/1–4, p. 39.

UGANDA. Bunyoro [Unyoro], *E. Brown* 407!; Teso District: Serere, Apr. 1932, *Chandler* 392a!
KENYA. W. Suk District: Kapenguria, 6 May 1953, *Padwa* 63!; Trans-Nzoia District: SE. of Kitale, Kiptoi Farm, 10 May 1971, *Mabberley* 1124!; N. Kavirondo District: Bungoma, 2 May 1944, *R. Cameron* in *Bally* 3167!
DISTR. **U** 2, 3; **K** 2, 3, 5; NE. Cameroon, Central African Republic, Sudan, Ethiopia and Somalia
HAB. As var. *barnimiana*; 1050–2200 m.

SYN. *Kosaria tropaeolifolia* Schweinf. in Verh. Zool.-Bot. Ges. Wien 18: 687 (1868)
Dorstenia tropaeolifolia (Schweinf.) Bureau in DC., Prodr. 17: 278 (1873): Engl., E.M. 1: 26, t. 9B (1898); Friis in Nordic Journ. Bot. 3: 536, fig. 1 (1983)

NOTE. Friis (1983) maintains var. *tropaeolifolia* as a distinct species, citing the constancy of features in cultivation and the correlation of leaf form with the colour of the receptacle, which is whitish tinged (rather than green to purplish brown), coupled with an overlapping range in similar habitats without the occurrence of intermediates. Further examination, however, shows that the features mentioned by Friis are not always constant. Moreover, collections with both peltate and basally attached leaves are found.

24. D. ellenbeckiana *Engl.* in E.J. 33: 116 (1902); Rendle in F.T.A. 6(2): 72 (1916); Friis in Nordic Journ. Bot. 3: 536, fig. 2 (1983). Type: Ethiopia, Arussi-Galla, near Burkar, *Ellenbeck* 2018 (B, holo.!)

Herb, tuberous; tuber discoid to subglobose, up to 4.5 cm. in diameter; stems (? seasonal), up to 1 cm. long, with short internodes. Leaves subrosulate; lamina papyraceous when dry (± bullate when fresh), broadly elliptic to oblong or broadly obovate to subobovate, 3.5–18 × 2.5–9.5 cm., apex rounded, base cordate to almost rounded, margin finely to coarsely dentate-crenate or crenate; both surfaces puberulous to hirtellous, especially on the veins and often also near the margin; lateral veins 5–8 pairs; petiole 1–8 cm. long, 2–4 mm. thick; stipules subovate to triangular or subulate, 1.5–3.5 mm. long, persistent. Inflorescences solitary; peduncle 6–14.5 cm. long, 2–3 mm. thick. Receptacle discoid, suborbicular, 1.5–2.5 cm. in diameter; flowering face circular, margin almost lacking; appendages in 2 rows, the inner marginal row with triangular to band-shaped or subspathulate appendages 0.5–3.5 mm. long, shorter and longer ones ± alternating, the outer (submarginal) row with ± 8–10 subulate to band-shaped appendages ± 3–12 mm. long pointing backwards. Staminate flowers crowded; perianth-lobes 2; stamens 2. Pistillate flowers numerous; perianth tubular; stigma 1. Endocarp-body ± tetrahedral, ± 2 mm. in diameter, ± tuberculate.

KENYA. Northern Frontier Province: Dandu, 6 May 1952, *Gillett* 13095! & Darecha, 5 May 1978, *Gilbert & Thulin* 1476!
DISTR. **K** 1; Ethiopia
HAB. *Acacia-Commiphora* bushland, on sandy or stony soil; 600–1200 m.

11. FICUS

L., Sp. Pl.: 1059 (1753) & Gen. Pl., ed. 5: 482 (1754)

Urostigma Gasp., Nova Gen. Fici: 7 (1844)
Galoglychia Gasp., Nova Gen. Fici: 10 (1844)
Sycomorus Gasp., Ricerche Caprif.: 86 (1845)
Pharmacosycea Miq. in Lond. Journ. Bot. 6: 525 (1847)

Trees, shrubs or sometimes lianas, terrestrial or hemi-epiphytic (and then with aerial roots, often strangling and secondarily terrestrial), monoecious or dioecious. Leaves in spirals, distichous, subopposite or subverticillate; lamina pinnately veined with glandular (sometimes waxy) spots in the axils of at least the basal lateral veins beneath or (mostly a single spot) at the base of the midrib beneath; stipules fully amplexicaul, semi-amplexicaul or lateral, free or partly connate. Inflorescences (figs; syconia) solitary or in pairs in the leaf-axils, mostly several together on small spurs in the leaf-axils down to the lesser branches or on distinct spurs on the lesser branches down to the trunk, or on leafless branchlets on the older wood down to the base of the trunk. Figs unisexual, containing staminate flowers, seed flowers (destined to produce seeds) and gall flowers (destined to hatch the larvae of the pollinator) or the figs on one plant containing staminate flowers and gall flowers (gall figs) and on another plant containing only seed flowers (seed figs); bracts 2–3(–4) on the peduncle (peduncular bracts) or subtending the receptacle (basal bracts), sometimes on the outer surface of the receptacle (lateral bracts), many in the opening of the receptacle (ostiolar bracts), and often among the flowers (interfloral bracts). Receptacle urceolate, with a narrow apical opening (ostiole). Staminate flowers near the ostiole or dispersed, mostly pedicellate; tepals 2–6, free or connate; stamens 1 or 1–3; pistillode present or absent. Pistillate flowers sessile or pedicellate; tepals 2–6(–7), free or connate; ovary free; stigmas 1, filiform or infundibuliform, or 2 and ± filiform; pistillate flowers ± strongly differentiated into seed flowers (sessile or shortly pedicellate, ovary often oblong to ovate and the style relatively long) and gall flowers (mostly pedicellate, ovary often ± obovate and the style relatively short, in dioecious species; gall flowers mostly not functional). Fruits achene-like or often ± drupaceous, releasing the endocarp-body (pyrene) or not; at fruit the wall of the fig mostly ± fleshy, coloured or green; gall fruits achene-like.

The genus comprises ± 750 species, ± 500 species in Asia and Australasia, ± 150 species in the Neotropics, and ± 100 species in Africa and vicinity.

The present subdivision of the genus was proposed by Corner in Gard. Bull. Singapore 21: 3–6 (1965), with full synonymy. The principal diagnostic features are shown in Fig. 16, p. 47.

KEY TO CULTIVATED SPECIES

The following species have all been introduced from Asia. Further information may be found in T.T.C.L. (1949), U.O.P.Z. (1949) and Dale, Introd. Trees Ug. (1952).

Root-climber; leaves heteromorphic on sterile and fertile branches; figs pear-shaped, up to 5(–7) cm. long . . . *F. pumila* L.
Trees or shrubs; leaves not heteromorphic:
  Lamina 3–5-lobed; figs ± broadly pear-shaped, pedunculate . . . *F. carica* L.
  Lamina entire; figs globose to ellipsoid, sessile:
    Lamina with 15 or more pairs of lateral veins; stipules long; figs ellipsoid . . . *F. elastica* Roxb.
    Lamina with at most 15 pairs of lateral veins; stipules short; figs mostly globose:
      Lamina with a truncate base and a caudate apex, margin repand; petiole long and slender . . . *F. religiosa* L.
      Lamina with a rounded to acute base and a ± rounded or acute apex, margin entire; petiole short or stout:
        Lamina usually longer than 10 cm.; petiole usually longer than 2 cm.; figs 1–1.5 cm. in diameter . . . *F. benghalensis* L.
        Lamina usually shorter than 10 cm.; petiole up to 2 cm. long; figs ± 0.5–1 cm. in diameter:
          Apex of the lamina distinctly acuminate; figs turning yellow, orange or red . . . *F. benjamina* L.
          Apex of the lamina shortly and faintly acuminate; figs turning purple or blackish . . . *F. microcarpa* L.f.

KEY TO SECTIONS OF NATIVE SPECIES

Ostiole circular, at least 3 ostiolar bracts visible, none or only the lower ostiolar bracts descending (fig. 16/1, 2, 6, 7, 9, 10):
  Several ostiolar bracts visible (fig. 16/9); glandular spot on the lamina beneath in the axils of the main basal lateral nerves (fig. 16/22):
    Trees or shrubs, dioecious; stipules semi-amplexicaul to lateral . . . I. sect. **Sycidium**
    Trees, monoecious; stipules fully amplexicaul . . . II. sect. **Sycomorus**
  Three ostiolar bracts visible (fig. 16/10); 1–2 glandular spots on the lamina beneath at the base of the midrib (fig. 16/23, 24):
    Lenticels concentrated on the uppermost part of the internodes; stigmas 2 (fig. 16/14) . . . III. sect. **Oreosycea**
    Lenticels scattered over the internodes; stigmas usually 1 (fig. 16/15) . . . IV. sect. **Urostigma**, p. 46
Ostiole slit-shaped (fig. 16/11); all ostiolar bracts descending (fig. 16/8) . . . V. sect. **Galoglychia**, p. 46

I. sect. **Sycidium** *Miq.*

Shrubs or trees, terrestrial, dioecious; sap watery. Leaves distichous to sometimes alternate, subopposite or verticillate, often lobed to divided especially when juvenile;

tertiary venation scalariform to reticulate; glandular spots in the axils of the main basal lateral veins beneath; stipules lateral to semi-amplexicaul, free. Figs often solitary in the leaf-axils or on minute spurs on the older wood, pedunculate, with 2–4 bracts on the peduncle (often not in a whorl), several lateral bracts; ostiole circular, with several ostiolar bracts visible, only the lower ostiolar bracts descending; interfloral bracts absent. Staminate flowers near the ostiole; tepals 3–6, at least ciliolate; stamens 1–3; pistillode present. Seed and gall flowers distinct; tepals 4–6, at least ciliolate; stigma 1; endocarp-body often released. Wall of the fruiting fig soft, red, orange or yellow. Species 1–3.

Lamina asymmetrical; lateral veins 3–10 pairs; peduncle up to 4 mm. long	2. *F. asperifolia*
Lamina symmetrical or if asymmetrical then the lateral veins 3–5 pairs; peduncle at least 5 mm. long:	
Lateral veins 3–5(–6) pairs; leaves mostly alternate; trees	1. *F. exasperata*
Lateral veins 5–12 pairs; leaves partly subopposite or subverticillate; shrubs	3. *F. capreifolia*

II. subgen. **Sycomorus** (*Gasp.*) *Mildbr. & Burret* sect. **Sycomorus**

Trees or less often shrubs, terrestrial, monoecious; sap milky. Leaves in spirals or tending to distichous, margin mostly dentate, crenate or repand; tertiary venation for the greater part scalariform; glandular spots in the axils of the main basal lateral veins; stipules fully amplexicaul, free. Figs solitary (or in pairs) in the leaf-axils or 1–3 together on unbranched or branched leafless branchlets on the older wood, often down to the trunk, pedunculate; basal bracts 3; lateral bracts absent; receptacle rather large; ostiole circular, with several ostiolar bracts visible, only the lower ones descending; interfloral bracts lacking among the pistillate flowers. Staminate flowers near the ostiole, initially enveloped by 2(–3) bracts; perianth tubular, 3-lobed; stamens 2–3. Seed and gall flowers rather distinct; tepals 2–4–6, irregularly shaped, free or connate; stigma 1; endocarp-body often released. Wall of the fruiting fig soft, red to orange or yellowish. Species 4–8.

Figs solitary in the leaf-axils (or just below the leaves):	
Lamina scabrous above; hairs on the petiole distinctly different in length	4. *F. sycomorus*
Lamina smooth above; hairs on the petiole not distinctly different in length	8. *F. vallis-choudae*
Figs on leafless branchlets on the older wood, often down to the trunk:	
Stipules subpersistent; basal bracts (fig. 16/2, 5) 3–5 mm. long; trees geocarpic	7. *F. vogeliana*
Stipules caducous; basal bracts 2–3 mm. long; trees not geocarpic:	
Hairs of leafy twigs and petiole conspicuously different in length; periderm of branchlets and petiole flaking off:	
Apex of lamina rounded to obtuse; petiole up to 4(–6) cm. long; trunk short	4. *F. sycomorus*
Apex of the lamina acuminate to acute; petiole up to 9 cm. long; trunk tall	5. *F. mucuso*
Hairs of leafy twigs and petiole not conspicuously different in length (or absent); periderm of branchlets and petiole usually not flaking off	6. *F. sur*

III. subgen. **Pharmacosycea** (*Miq.*). *Miq.* sect. **Oreosycea** (*Miq.*) *Miq.*

Trees or less often shrubs, terrestrial, monoecious; sap scanty and watery; lenticels concentrated on the uppermost part of the internodes. Leaves in spirals or distichous, margin entire to dentate, when juvenile irregularly pinnately incised; tertiary venation for the greater part scalariform; glandular spots 1–2 at the base of the midrib beneath; stipules fully amplexicaul, when juvenile semi-amplexicaul, free. Figs solitary or in pairs in the leaf-axils, pedunculate; basal bracts 3; lateral bracts occasionally present; ostiole circular with 3 ostiolar bracts visible, only the lower ostiolar bracts descending; interfloral bracts lacking in the two African species. Staminate flowers near the ostiole; tepals 2–3, connate, glabrous; stamen 1; pistillode often present. Seed and gall flowers rather distinct; tepals 2–3, connate, glabrous; stigmas 2, ± filiform; fruits achene-like. Wall of the fruiting receptacle firm, yellow to orange. Species 9,10.

Lateral veins (5–)8–13 pairs; lamina often smooth above . . . 9. *F. variifolia*
Lateral veins 5–8(–9) pairs; lamina scabrous above . . . 10. *F. dicranostyla*

IV. subgen. **Urostigma** (*Gasp.*) *Miq.* sect. **Urostigma**

Trees or shrubs, usually terrestrial (or epilithic), monoecious; sap milky. Leaves in spirals, margin entire or almost so; tertiary venation reticulate; glandular spot at the base of the midrib beneath; stipules fully amplexicaul, free. Figs in pairs or sometimes more together in the leaf-axils or just below the leaves, pedunculate or sessile; basal bracts 3; lateral bracts absent; ostiole circular, with 3 ostiolar bracts visible, only the lower ostiolar bracts descending; interfloral bracts absent or few. Staminate flowers usually only near the ostiole; tepals 3–4, free; stamen 1; pistillode sometimes present. Seed and gall flowers ± distinct; tepals 3–4(–7), free; stigma 1, ± filiform; fruit achene-like. Wall of the fruiting fig soft, red, pink or purplish. Species 11–13.

Basal lateral veins usually branched, almost straight (not running parallel to the leaf-margin), the other lateral veins often dividing far from the margin . . . . . 11. *F. ingens*
Basal lateral veins unbranched, ± curved, running almost parallel to the leaf-margin; the other lateral veins dividing not far from the margin:
  Lamina ± 5(–8) times as long as the petiole; stipules up to 1.5 cm. long; figs mostly smooth when dry; tree of dry places 12. *F. cordata*
  Lamina ± (5–)8–10 times as long as the petiole; stipules up to 4 cm. long; figs wrinkled when dry; shrub or treelet of wet places . . . . 13. *F. verruculosa*

V. subgen. **Urostigma** (*Gasp.*) *Miq.* sect. **Galoglychia** (*Gasp.*) *Endl.*; C.C. Berg in Konink. Nederl. Akad. Weten., Ser. C, 89:121–127 (1986)

subgen. *Bibracteatae* Mildbr. & Burret in E.J. 46: 175 (1911)

Trees, shrubs or sometimes lianas, hemi-epiphytic or terrestrial (or epilithic), mostly with aerial roots; sap milky. Leaves in spirals, sometimes almost distichous and/or

---

FIG. 16. Schematic diagrams of key features in *Ficus* **1**, receptacle with peduncular, lateral, and ostiolar bracts (sect. *Sycidium);* **2**, receptacle with basal and ostiolar bracts (other sections); **3**, stipitate receptacle; **4**, basal bracts two (sect. *Galoglychia*); **5**, basal bracts three (subgen. *Sycomorus*, sect. *Oreosycea*, sect. *Urostigma*); **6**, most ostiolar bracts interlocking (sect. *Sycidium*, subgen. *Sycomorus*, sect. *Urostigma*); **7**, the middle ostiolar bracts not interlocking (sect. *Oreosycea*); **8**, all ostiolar bracts descending (sect. *Galoglychia*); **9**, ostiole circular, several ostiolar bracts visible (sect. *Sycidium*, subgen. *Sycomorus*); **10**, ostiole circular, three ostiolar bracts visible (sect. *Oreosycea*, sect. *Urostigma*); **11**, ostiole slit-shaped (sect. *Galoglychia*); **12**, stigma ± infundibuliform (sect. *Sycidium*, subgen. *Sycomorus*); **13**, stigma ± flame-shaped and slightly infundibuliform (subgen. *Sycomorus*); **14**, stigmas two (sect. *Oreosycea*, and sometimes in sect. *Galoglychia*); **15**, stigma one, elongate (sect. *Urostigma*, sect. *Galoglychia*); **16**, long-styled ('seed') flower, often sessile or short-pedicellate; **17**, short-styled ('gall') flower, (± long-pedicellate); **18**, staminate flower with distinct pistillode and hairy tepals (sect. *Sycidium*); **19**, staminate flower with saccate perianth and two stamens, being enveloped by two large bracts or bracteoles (subgen. *Sycomorus*); **20**, staminate flower with one stamen (sect. *Oreosycea*, sect. *Urostigma*, sect. *Galoglychia*); **21**, calyptrate bud-cover (as found in, e.g., *F. craterostoma* and *F. ovata*); 21a, diagram of same; **22**, glandular spots in the axils of the basal (or main) lateral veins beneath (sect. *Sycidium*, subgen. *Sycomorus*); **23**, glandular spots two, on the base of the midrib beneath (sect. *Oreosycea*); **24**, glandular spot one, on the base of the midrib beneath (sect. *Urostigma*, sect. *Galoglychia*); **25**, basal lateral veins branched; **26**, basal lateral veins unbranched; **27**, tertiary venation scalariform; **28**, tertiary venation predominantly parallel to the lateral veins; **29**, tertiary venation reticulate. Si = sect. *Sycidium*; So = subgen. *Sycomorus*; O = sect. *Oreosycea*; U = sect. *Urostigma*; G = sect. *Galoglychia*.

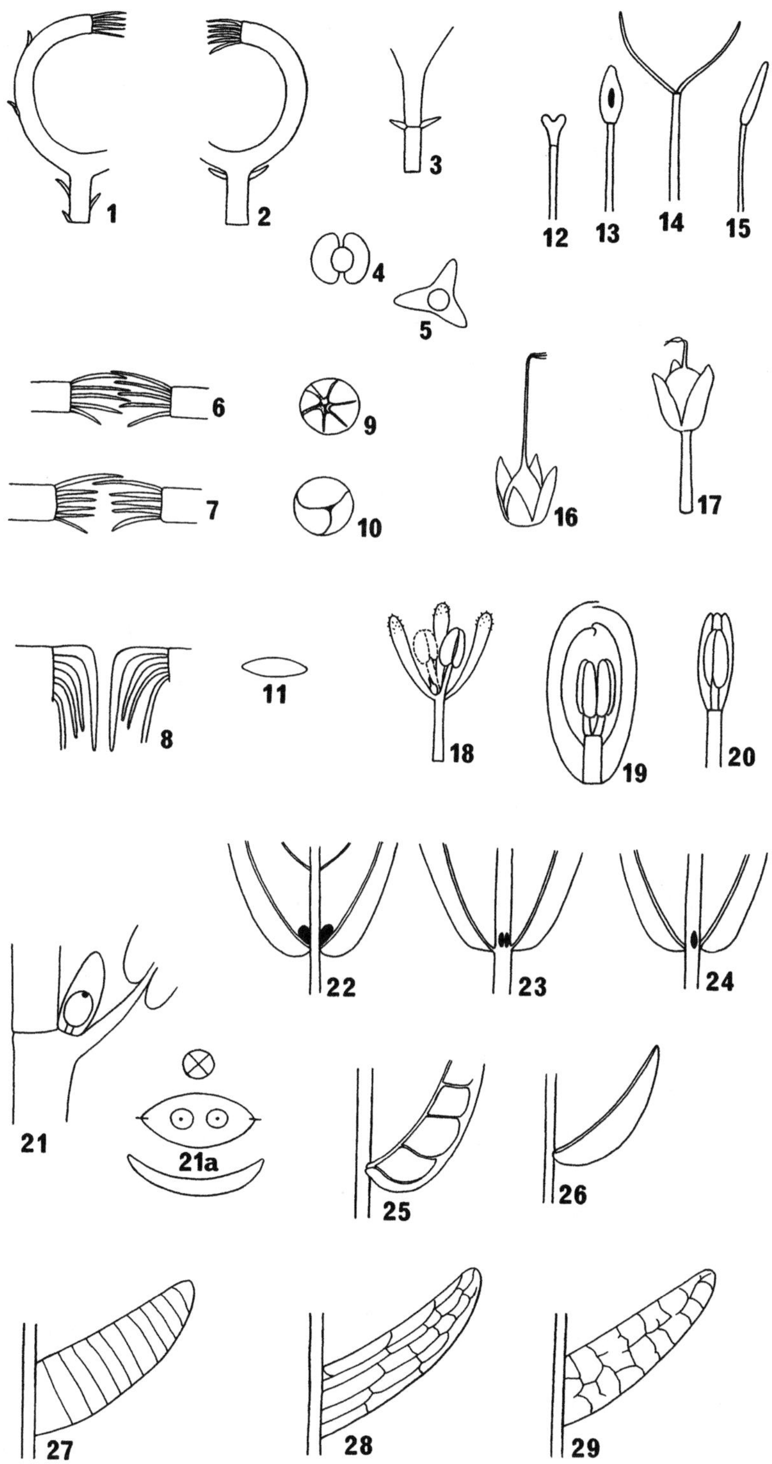
1
2
3
4
5
6
7
8
9
10
11
12
13
14
15
16
17
18
19
20
21
21a
22
23
24
25
26
27
28
29

subopposite, margin entire or almost so; tertiary venation scalariform to reticulate or ± parallel to the lateral veins; glandular spot at the base of the midrib beneath; stipules fully amplexicaul, free or partly connate. Figs in the leaf-axils or just below the leaves or on spurs on the lesser branches or down to the base of the trunk, pedunculate or sessile; basal bracts 2(–3); lateral bracts absent; ostiole slit-shaped or triradiate, all ostiolar bracts descending; interfloral bracts present. Staminate flowers dispersed; tepals 2–4, free or connate, glabrous; stamen 1; pistillode usually absent. Seed and gall flowers ± different; tepals 2–4, free or partly connate, glabrous; stigmas 1(–2); fruit achene-like or drupaceous and then often releasing the endocarp-body or the upper part forming a mucilaginous cap. Wall of the fruiting fig soft to rather firm, reddish, orange, yellow, greenish, purplish or brownish. Species 14–50.

Within this section 6 ± distinct groups (subsections) can be recognized, see C.C. Berg (1986): species 14–23; 24, 25; 26–32; 33–40; 41–43; 44–50.

KEY TO ARTIFICIAL GROUPS OF SECT. GALOGLYCHIA

Figs with a peduncle at least 2 mm. long:
  Figs on distinct spurs (0.5–5(–15) cm. long) down to the main branches or trunk; receptacle at least 1 cm. in diameter when dry . . . . **Group 1**
  Figs not on spurs, or on minute 1–2 mm. long spurs in the leaf-axils and/or just below the leaves and then figs at most 1 cm. in diameter when dry . . . . **Group 2**
Figs sessile or subsessile:
  Stipules and/or part of the calyptrate bud-cover initially enclosing the figs subpersistent . . . . **Group 3,** p.51
  Stipules and part of the calyptrate bud-cover initially enclosing the figs (if present) caducous . . . . . **Group 4,** p.51

Group 1. Figs pedunculate on spurs down to the main branches or trunk, at least 1 cm. in diameter when dry.

Basal bracts (fig. 16/2, 4) caducous:
  Lateral veins 10–16 pairs; lamina usually broadest above the middle; petiole 1–2.5(–3.5) cm. long; midrib and petiole drying pale brown or grey . . . . 35. *F. artocarpoides*
  Lateral veins 5–10 pairs, if up to 14 pairs, then lamina broadest below the middle, the petiole up to 5.5(–8) cm. long and/or the midrib and petiole drying red-brown:
    Stipules with short white appressed hairs (but not ciliolate); lamina usually minutely puberulous on the main veins beneath; figs subglobose, 1.5–3 cm. in diameter when dry (2.5–4 cm. when fresh) . . . 36. *F. chirindensis*
    Stipules glabrous, with minute patent hairs or distinctly ciliolate; lamina (except sometimes midrib) glabrous beneath:
      Figs globose, when dry 1.5–3.5 cm. in diameter, mostly wrinkled . . . . 37. *F. sansibarica*
      Figs ellipsoid, if globose, when dry at most 1.5 cm. in diameter, mostly smooth:
        Petiole when dry (1–)1.5–2 mm. thick; peduncle when dry 1–1.5 mm. thick, if 0.5–1 mm. thick, then base of the lamina acute to obtuse or the lamina 8–16 cm. long . . . . 33. *F. ottoniifolia*
        Petiole when dry 0.5–1 mm. thick; peduncle when dry 0.5–1 mm. thick; base of the lamina rounded to cordulate; lamina 2–9 cm. long . . . . . 34. *F. tremula*
Basal bracts persistent:
  Stipules with short white appressed hairs; lamina usually minutely puberulous on the main veins beneath; figs subglobose, 1.5–3 cm. in diameter when dry (2.5–4 cm. when fresh) . . . . 36. *F. chirindensis*

Stipules glabrous (or with minute patent hairs); lamina (except sometimes midrib) glabrous beneath:
  Figs ellipsoid, if globose, then when dry at most 1.5 cm. in diameter, mostly smooth:
    Lamina 8–16 cm. long, base acute to obtuse . . . 33. *F. ottoniifolia*
    Lamina 2–9 cm. long, base cordulate to rarely rounded 34. *F. tremula*
  Figs globose, when dry 1.5–3 cm. in diameter, mostly wrinkled:
    Peduncle 0.5–1(–1.1) cm. long; petiole when dry 2–4 mm. thick; apex of the thickly coriaceous lamina rounded or very shortly and bluntly acuminate 39. *F. bubu*
    Peduncle (0.9–)1–2 cm. long; petiole when dry 1–2(–3) mm. thick; apex of the coriaceous to subcoriaceous lamina usually acuminate:
      Spurs up to 5(–15) cm. long, their bud-scales ± densely puberulous: midrib and petiole usually drying red-brown . . . . 37. *F. sansibarica*
      Spurs up to 1 cm. long, their bud-scales glabrous; midrib and petiole usually drying blackish 38. *F. polita*

Group 2. Figs pedunculate in the leaf-axils or if just below then only up to 1 cm. in diameter when dry.

Stipules persistent or subpersistent:
  Stipules basally connate, 1.5–3 cm. long; figs when dry at least 1 cm. in diameter:
    Peduncle 0.5–2.5 cm. long; wall of the figs spongy, when dry 2–5 mm. thick . . . . 46. *F. cyathistipula*
    Peduncle up to 0.4 cm. long; wall of the figs not spongy, when dry 0.5–1 mm. thick . . . . 47. *F. densistipulata*
  Stipules free, 0.5–1.5 cm. long; figs when dry at most 1 cm. in diameter:
    Petiole 0.2–1(–1.5) cm. long; lamina glabrous; midrib not reaching the apex of the lamina:
      Lamina usually obovate to elliptic; petiole when dry 1–1.5 mm. thick; petiole (0.3–)0.5–1.5 cm. long; figs when dry 0.4–0.8(–1.4) cm. in diameter . . . 30. *F. faulkneriana*
      Lamina usually subobovate to oblong or oblanceolate; petiole when dry 0.5–1 mm. thick; peduncle 0.2–0.5 cm. long; figs when dry 0.3–0.4 cm. in diameter 31. *F. lingua*
    Petiole 1–4(–6) cm. long, if shorter than 1 cm. then the lamina ± hairy, the midrib reaching the apex of the lamina and/or the figs hairy . . . . 32. *F. thonningii*
Stipules caducous:
  Tertiary venation partly perpendicular to the lateral veins, basal pair of lateral veins branched (base of the lamina often cordate and/or the lower surface hairy):
    Epiderm of the petiole flaking off when dry . . . . 16. *F. vasta*
    Epiderm of the petiole not flaking off:
      Basal bracts caducous:
        Lamina hairy beneath, at least in the axils of the lateral veins: petiole when dry 2–4 mm. thick 21. *F. abutilifolia*
        Lamina glabrous beneath; petiole when dry 1–2 mm. thick . . . . 22. *F. populifolia*
      Basal bracts persistent:
        Figs initially enclosed by an up to 1.5 cm. long bud-cover; figs ellipsoid, if subglobose, then when dry at least 1 cm. in diameter and/or on peduncles up to 0.5 cm. long . . . . . . 40. *F. ovata*

Figs initially not enclosed by a conspicuous bud-cover; figs subglobose, if ellipsoid, then when dry at most 1 cm. in diameter or the peduncle 1–2.5 cm. long:
Lateral veins 3–7 pairs . . . . 18. *F. glumosa*
Lateral veins 7–16 pairs:
Lateral veins 7–11 pairs; peduncle 0.5–1 cm. long; periderm of the leafy twigs not flaking off; stipules 1.5–4.5(–8) cm. long . . . 23. *F. trichopoda*
Lateral veins 10–16 pairs; peduncle 1–2.5 cm. long; periderm at least of the older parts of the leafy twigs, flaking off or the stipules up to 1.2 cm. long:
Stipules hairy, 0.8–3.5 cm. long . . . . . 14. *F. platyphylla*
Stipules glabrous or only hairy at the base, 0.8–1.2 cm. long . . . . 15. *F. bussei*
Tertiary venation reticulate or partly parallel to the lateral veins, basal lateral veins mostly unbranched (base of the lamina usually acute to rounded and/or the lower surface glabrous):
Figs 2–7 together on short spurs in the leaf-axils or just below the leaves:
Lateral veins 14–27 pairs; peduncle 0.3–0.8 cm. long 41. *F. pseudomangifera*
Lateral veins 10–13 pairs; peduncle 0.8–1.7 cm. long 42. *F. usambarensis*
Figs solitary or in pairs in the leaf-axils and/or just below the leaves:
Basal bracts caducous:
Figs 0.3–0.4 cm. in diameter when dry; lamina 0.5–5 cm. long; petiole 0.2–0.8 cm. long . . . . 31 *F. lingua*
Figs 0.5–2 cm. in diameter when dry; lamina mostly more than 5 cm. long and petiole mostly longer than 1 cm.:
Midrib reaching the apex of the lamina; apex of the lamina acuminate to acute; lamina mostly 10–20 cm. long:
Figs 1.5–2 cm. in diameter when dry; lamina ovate to elliptic . . . . 26. *F. fischeri*
Figs 0.5–1 cm. in diameter when dry; lamina usually lanceolate to linear or oblong 44. *F. barteri*
Midrib not reaching the apex of the lamina; apex of the lamina bluntly and shortly acuminate to rounded or emarginate; lamina mostly 2.5–9 cm. long . . . . 29. *F. natalensis*
Basal bracts persistent:
Figs 1–2(–2.5) cm. in diameter when dry:
Lateral veins 3–7 pairs; basal lateral veins usually branched . . . . 18. *F. glumosa*
Lateral veins (5–)7–18 pairs; basal lateral veins unbranched:
Petiole 2–3 mm. thick; lamina mostly 10–20 cm. long; figs 3–4.5 cm. in diameter when fresh 48. *F. scassellatii*
Petiole 1–2 mm. thick; lamina mostly 5–10 cm. long; figs at most 2 cm. in diameter when fresh 32. *F. thonningii*
Figs at most 1 cm. in diameter when dry; petiole 1–2(–2.5) mm. thick:
Figs 0.3–0.4 cm. in diameter when dry; peduncle up to 0.5 cm. long; petiole 2–5 mm. long, 0.5–1 mm. thick . . . . 31. *F. lingua*
Figs 0.4–1 cm. in diameter when dry; peduncle more than 0.5 cm. long; petiole longer than 8 mm. and/or 1–2(–2.5) mm. thick:

Peduncle at most 3 mm. long:
Periderm of older parts of the leafy twigs flaking off; smaller veins inconspicuous — 18. *F. glumosa*
Periderm of the leafy twigs not flaking off; smaller veins ± conspicuous . . . . — 32. *F. thonningii*
Peduncle at least 3 mm. long:
Petiole 0.5–1(–1.5) cm. long; lateral veins 3–8 pairs; midrib not reaching the apex of the lamina; stipules, figs and lamina glabrous or almost so — 30. *F. faulkneriana*
Petiole at least 1.5 cm. long, if shorter then lateral veins more than 8 pairs, midrib reaching the apex of the lamina and/or the stipules, figs and/or lamina hairy — 32. *F. thonningii*

Group 3. Figs sessile or subsessile; stipules and/or parts of the calyptrate bud-cover initially enclosing the figs subpersistent.

Tertiary venation partly scalariform; lamina usually hairy beneath . . . . — 25. *F. saussureana*
Tertiary venation reticulate or partly parallel to the lateral veins; lamina usually glabrous beneath:
Petiole ± 4 mm. thick when dry; basal bracts ± 5 mm. long; lamina mostly 20–30 cm. long . . . . — 49. *F. preussii*
Petiole up to 2.5 mm. thick when dry, if up to 4 mm. thick then basal bracts at most 3.5 mm. long and lamina mostly 10–20 cm. long:
Stipules basally connate, 1.5–3 cm. long:
Lateral veins 8–11 pairs; figs not stipitate . . . . — 45. *F. conraui*
Lateral veins 6–8 pairs; figs stipitate . . . . . . — 47. *F. densistipulata*
Stipules free, 0.5–1.5 cm. long:
Figs 1.2–2 cm. in diameter when dry; stipules 0.5–1.5 cm. long, glabrous outside, pubescent inside at least at the base . . . . — 27. *F. amadiensis*
Figs at most 1 cm. in diameter when dry; stipules up to 0.5 cm. long, if longer then outside puberulous or pubescent and inside glabrous:
Figs 0.3–0.4 cm. in diameter when dry . . . . . — 31. *F. lingua*
Figs 0.5–1 cm. in diameter when dry:
Apex of the lamina truncate, emarginate or 2-lobed; leaves often subopposite; petioles ± equally long . . . . — 28. *F. craterostoma*
Apex of the lamina acuminate to rounded; midrib often reaching the apex of the lamina; leaves rarely subopposite; petioles on the same twig often different in length — 32. *F. thonningi*

Group 4. Figs sessile or subsessile; stipules and bud-cover (if present) caducous.

Epiderm of the petiole flaking off when dry:
Main basal lateral veins reaching the margin in or above the middle of the lamina; lamina often almost as wide as long:
Stipules 2–5 cm. long, on flush up to 8.5 cm. long; basal bracts caducous . . . . — 16. *F. vasta*
Stipules 0.5–1.5 cm. long, on flush up to 4 cm. long; basal bracts persistent . . . . — 17. *F. wakefieldii*
Main basal lateral veins reaching the margin far below the middle of the lamina; lamina much longer than wide:

Lateral veins (4–)6–10(–12) pairs	24. *F. lutea*
Lateral veins 12–20 pairs	25. *F. saussureana*
Epiderm of the petiole not flaking off:	
Lamina glabrous beneath:	
Figs 1–4 cm. in diameter when dry:	
Stipules ± 1.5 cm. long; petiole 1.5–2 mm. thick when dry	43. *F. oreodryadum*
Stipules 0.3–1 cm. long; petiole 4–10 mm. thick when dry:	
Tertiary venation partly scalariform; petiole 2–4 mm. thick when dry; lamina mostly 10–30 cm. long	40. *F. ovata*
Tertiary venation reticulate; petiole 4–10 mm. thick when dry; lamina mostly 25–60 cm. long	50. *F. wildemaniana*
Figs at most 1 cm. in diameter when dry:	
Base of the lamina cordate to rounded:	
Lateral veins 3–7 pairs; periderm of the leafy twigs flaking off at least the older parts	18. *F. glumosa*
Lateral veins (5–)7–12(–16) pairs; periderm of the leafy twigs not flaking off	32. *F. thonningii*
Base of the lamina cuneate to obtuse:	
Figs 0.3–0.4 cm. in diameter; petiole 0.2–0.8 cm. long, 0.5–1 mm. thick	31. *F. lingua*
Figs 0.5–1 cm. in diameter when dry; petiole 0.5–4(–6) cm. long, 1–2 mm. thick:	
Midrib not reaching the apex of the lamina; apex of the lamina truncate, emarginate or 2-lobed; leaves often subopposite	28. *F. craterostoma*
Midrib often reaching the apex of the lamina; apex of the lamina acuminate to rounded; leaves rarely subopposite	32. *F. thonningii*
Lamina hairy at least beneath:	
Lateral veins (5–)7–12(–16) pairs; basal lateral veins unbranched	32. *F. thonningii*
Lateral veins 3–7 pairs; basal lateral veins branched (at least faintly so):	
Petiole 0.5–1 mm. thick when dry; periderm of the leafy twigs not flaking off	20. *F. nigro-punctata*
Petiole 1–2 mm. thick when dry; periderm of the leafy twigs flaking off at least the older parts:	
Receptacle 0.5–1 cm. in diameter when dry; parts of the indumentum yellowish; vein-reticulum of leaves plane beneath, inconspicuous . . .	18. *F. glumosa*
Receptacle (0.7–)1–1.8 cm. in diameter when dry; indumentum whitish; vein-reticulum of leaves prominent beneath, conspicuous	19. *F. stuhlmannii*

1. F. exasperata *Vahl*, Enum. Pl. 2: 197 (1805); Hutch. in F.T.A. 6(2): 110 (1916); Peter, F.D.O.-A. 2: 94 (1932); Lebrun & Boutique in F.C.B. 1: 126 (1948); T.T.C.L.: 353 (1949); U.O.P.Z.: 262 (1949); I.T.U., ed. 2: 248 (1952); F.P.S. 2: 268, fig. 95 (1952); F.W.T.A., ed. 2, 1: 605, fig. 173 (1958); K.T.S.: 316 (1961); F.F.N.R.: 30 (1962); Aweke in Meded. Landb. Wageningen 79-3: 21, fig. 5 (1979); Hamilton, Ug. For. Trees: 96 (1981); Troupin, Fl. Pl. Lign. Rwanda: 444, fig. 147.6 (1982); C.C. Berg et al. in Fl. Cameroun 28: 121, t. 39 (1985). Type: Ghana, *Isert* (C, holo.!, B, iso.!)

Shrub or tree up to 20(–30) m. tall. Leafy twigs 1–5 mm. thick, hispidulous. Leaves almost distichous and alternate, sometimes subopposite; lamina subcoriaceous to coriaceous, ovate to elliptic or obovate, sometimes oblong or subcircular, 2.5–16 × 1–12 cm., sometimes ± asymmetrical, apex shortly acuminate, sometimes acute, obtuse or rounded, base acute to obtuse or occasionally subcordate, margin dentate to subentire; upper surface scabrous, hispidulous, lower surface scabrous, hispidulous or partly hirtellous to subtomentose; lateral veins 3–5(–6) pairs; petiole 0.5–2.5 cm. long; stipules 0.2–0.5 cm. long, strigillose to strigose, caducous. Figs in pairs or solitary in the leaf-axils, just below the leaves or sometimes on the older wood; peduncle 0.5–1(–1.5) cm. long; 1–5 broadly ovate bracts scattered on the peduncle, 1–4 similar ones on the outer surface of the receptacle. Receptacle subglobose, 1–2.5 cm. in diameter when fresh, 0.8–1.5 cm. when dry, hispidulous, yellow, orange or reddish at maturity.

UGANDA. Mengo District: Busiro, *Dawe* 112! & Ntakafunvu, Dec. 1913, *Dummer* 533! & Old Entebbe, Feb. 1932, *Eggeling* 184!

KENYA. Teita District: below Wusi, 7 Feb. 1953, *Bally* 8773!; Kwale District: Diani Beach, 9 July 1968, *Gillett* 18636!; Kilifi District: Chasimba, 17 Nov. 1974, *B.R. Adams* 102!

TANZANIA. Lushoto District: Soni, 12 Mar. 1972, *Faulkner* 4691!; Mpanda District: Sisaga, 22 Sept. 1958, *Newbould & Jefford* 2553!; Morogoro, Aug. 1952, *Semsei* 854!; Zanzibar I., Chwaka, 6 Nov. 1959, *Faulkner* 2393!

DISTR. **U** 4; **K** 3–5, 7; **T** 2–8; **Z**; **P**; extending to Senegal, Djibouti, Angola, Mozambique and E. Zimbabwe, also in Yemen, Sri Lanka and S. India

HAB. Forest, often at edges, in rocky places and along rivers, sometimes persisting in cleared places; 0–1200(–1850) m.

2. F. asperifolia *Miq.* in Lond. Journ. Bot. 7: 231, 564, fig. 15B (1848); Hook. f., Niger Fl.: 524 (1849); Hutch. in F.T.A. 6(2): 111 (1916); Lebrun & Boutique in F.C.B. 1: 127 (1948); F.W.T.A., ed. 2, 1: 606 (1958); F.F.N.R.: 29 (1962); Troupin, Fl. Pl. Lign. Rwanda: 440, fig. 148.1 (1982); C.C. Berg et al. in Fl. Cameroun 28: 124, t. 40 (1985). Type: Nigeria, R. Niger, Aboh, *Vogel* 47 (K, holo.!)

Shrub up to 5 m. tall, often with whippy, straggling or subscandent branches. Leafy twigs 1–5 mm. thick, white to brown hirtellous, hispidulous ± strigillose or almost glabrous. Leaves distichous, alternate; lamina chartaceous to subcoriaceous, elliptic to oblong, ovate, subobovate, lanceolate or sometimes linear, 3–12 × 1.5–12 cm., usually ± asymmetrical, apex acuminate to caudate, acute or sometimes obtuse, base acute to obtuse or rounded, margin dentate to irregularly pinnately lobed or divided or sometimes subentire; upper surface scabrous, hispidulous or strigillose, lower surface hispid to hirtellous, sometimes almost glabrous; lateral veins 3–10 pairs, in large leaves up to 13, or in very narrow leaves often more than 13; petiole 0.5–2 cm. long; stipules 0.3–0.6 cm. long, puberulous to almost glabrous, caducous. Figs 1–3(–5) together in the leaf-axils or just below the leaves, sometimes on the older wood; peduncle up to 4 mm. long; 2–4 small bracts scattered on the peduncle, 2–4 similar bracts on the outer surface of the receptacle. Receptacle depressed globose to globose or obovoid, ± 1–2 cm. in diameter when fresh, 0.5–1.2 cm. when dry, hispid or hispidulous, dark red to orange or yellowish at maturity.

UGANDA. Toro District: Butiti Hill, Sept 1936, *Mukibi* in *A.S. Thomas* 2611!; Masaka District: Kalungu county, 0.5 km. S. of W. Mengo border, 5 June 1971, *Lye* 6189!; Mengo District: Old Entebbe, Nov. 1931, *Eggeling* 127!

KENYA. Trans-Nzoia District: Kiminini, July 1966, *Tweedie* 3305!; S. Kavirondo District: Kisii, Sept. 1932, *Napier* 5275!; Kericho District: Ngoina Tea Estate, 13 Dec. 1967, *Perdue & Kibuwa* 9347!

TANZANIA. Bukoba District: Minziro Hill, 14 Aug. 1954, *Benedicto* 16!; Buha District: Kakombe valley, 7 Jan. 1964, *Pirozynski* 181!; Mpanda District: Mahali Mts., Katimba, 5 Sept. 1958, *Newbould & Jefford* 2349!

DISTR. **U** 1, 2, 4; **K** 3–5; **T** 1, 4; extending to Senegal, S. Sudan, N. Angola and N. Zambia

HAB. Forest, often at edges or riverine, and in wooded grassland; 650–1850 m.

SYN. *F. urceolaris* Hiern, Cat. Afr. Pl. Welw. 4: 1010 (1900); Peter, F.D.O.-A. 2: 96 (1932); Lebrun & Boutique in F.C.B. 1: 128 (1948); T.T.C.L.: 353 (1949); I.T.U., ed. 2: 261 (1952); F.P.S. 2: 268 (1952); K.T.S.: 321 (1961); Hamilton, Ug. For. Trees: 98 (1981). Type: Angola, Golungo Alto, *Welwitsch* 6390 (BM, lecto.!)

F. storthophylla Warb. in Ann. Mus. Congo, Bot., sér 6, 1: 32 (1904); Peter, F.D.O.-A. 2: 96 (1932); Lebrun & Boutique in F.C.B. 1: 128 (1948); T.T.C.L.: 353 (1949); K.T.S.: 320 (1961). Type: Uganda, Toro District, Yeria, *Scott Elliot* 7760 (B, holo.!, K, iso.!)

NOTE. The species is extremely variable in the shape and dimensions of the leaves. The material in the eastern part of the range differs from the material occurring more westwards in the smaller and shortly pedunculate to subsessile figs, but clear discontinuities in the variation have not been found.

3. F. capreifolia *Del.* in Ann. Sci. Nat., sér 2, 20: 94 (1843), as "*capreaefolia*"; Hutch. in F.T.A. 6(2): 107 (1916); Peter, F.D.O.-A. 2: 96 (1932); Lebrun & Boutique in F.C.B. 1: 126 (1948); T.T.C.L.: 353 (1949); I.T.U., ed. 2: 243 (1952); F.P.S. 2: 268 (1952); F.W.T.A., ed. 2, 1: 605 (1958); K.T.S.: 315 (1961); F.F.N.R..: 29, fig. 6A (1962); Aweke in Meded. Landb. Wageningen 79-3: 11, fig. 2 (1979); C.C. Berg et al., Fl. Cameroun 28: 127, t. 41 (1985). Type: Ethiopia, Tacazze R., *Galinier* 167 (MPU, holo.!)

Shrub up to 6 m. tall, usually with stiff branches. Leafy twigs 1–5 mm. thick, puberulous to hirsute. Leaves ± distinctly distichous and alternate or subopposite or the leaves subverticillate; lamina chartaceous, subovate to oblong or lanceolate, 2–15 × 1–5.5 cm., symmetrical, apex acute, obtuse, rounded, 3-lobed or 3-dentate, base rounded to obtuse or acute, margin subentire or ± faintly crenate; both surfaces scabrous; lateral veins 5–12 pairs; petiole 0.2–1(–2.5) cm. long; stipules 0.5–1 cm. long, partly puberulous, mostly subpersistent. Figs solitary or in pairs in the leaf-axils; peduncle 0.5–1.5 cm. long; 3 small bracts scattered on the peduncle or in a whorl, without bracts on the outer surface of the receptacle. Receptacle with a stipe up to 5 mm. long at least when dry, globose and 1.5–3 cm. in diameter when fresh, often pyriform and 1–2.5 cm. in diameter when dry, hispidulous, green to pale yellow at maturity. Fig. 17.

UGANDA. Acholi District: Gulu, Unyama R., July 1937, *Eggeling* 3354! & Agoro, Apr. 1938, *Eggeling* 3574!; Mbale District: NE. Sebei, Kirik R., Kakomei, Dec. 1942, *Dale* U.347!
KENYA. Northern Frontier Province: Kerio R., Lokori, 2 Sept. 1969, *Mwangangi* 1485! & Tana R., Kasha, 18 Nov. 1947, *J. Adamson* 444!; Embu District: Thiba R., near Mashamba, 29 Oct. 1973, *S.A. Robertson* 1891!
TANZANIA. Lushoto District: Pangani R., Korogwe, 6 Mar. 1954, *Faulkner* 1371!; Kondoa District: Bubu valley, 8 Jan. 1928, *B.D. Burtt* 1124!; Iringa District: Great Ruaha R., 35 km. from Msembe [Nsembi], 2 Dec. 1970, *Greenway & Kanuri* 14701!; Zanzibar I., Mahonda, 15 June 1962, *Faulkner* 3052!
DISTR. **U** 1, 3; **K** 1, 2, 4, 5, 7; **T** 1–7; **Z**; extending to Senegal, Ethiopia, Somalia, Angola and South Africa
HAB. Coastal, or riverine in low rainfall areas inland, often forming stands or thickets; 0–1450 m.

4. F. sycomorus *L.*, Sp. Pl.: 1059 (1753); Mildbr. & Burret in E.J. 46: 191 (1911); Hutch. in F.T.A. 6(2): 95 (1916); Peter, F.D.O.-A. 2: 90 (1932); T.T.C.L.: 355 (1949); I.T.U., ed. 2: 260 (1952); F.P.S. 2: 263, fig. 92 (1952); K.T.S.: 321 (1961); F.F.N.R.: 30, t. 6B (1962); Aweke in Meded. Landb. Wageningen 79-3: 72, fig. 18 (1979); Troupin, Fl. Pl. Lign. Rwanda: 449 (1982). Type: Egypt, not found (not in LINN)

Tree up to 20(–30) m. tall; trunk short; main branches spreading. Leafy twigs (1–)2–6 mm. thick, densely minutely puberulous and with much longer white to yellowish hairs especially on the nodes; periderm flaking off when dry. Lamina chartaceous to coriaceous, ovate to elliptic, obovate or subcircular, (1–)2.5–12(–21) × (0.5–)2–11(–16) cm., apex rounded to obtuse, base cordate to sometimes obtuse, margin subentire, slightly repand or denticulate; upper surface scabrous to scabridulous, sometimes almost smooth, hispidulous to strigillose, on the main veins whitish hirtellous to hirsute, lower surface puberulous to hispidulous, on the main veins partly whitish hirtellous or hirsute; lateral veins 5–10 pairs; petiole (0.5–)1–4(–6) cm. long, 1–3 mm. thick, densely minutely white puberulous and with much longer white to yellowish hairs, with the periderm flaking off when dry; stipules 0.5–2.5 cm. long, white puberulous to tomentose or partly hirtellous to hirsute, caducous. Figs solitary or sometimes in pairs in the leaf-axils or just below the leaves, on up to 10 cm. long unbranched leafless branchlets or on up to 20(–35) cm. long branched leafless branchlets on the older branches down to the trunk; peduncle 0.3–2.5 cm. long, 1–3 mm. thick; basal bracts 2–3 mm. long. Receptacle obovoid to pyriform or subglobose, often stipitate at least when dry, 1.5–5 cm. in diameter when fresh, (1–)1.5–3 cm. when dry, white to yellowish or brownish velutinous or densely tomentose to sparsely puberulous or pubescent, sometimes almost glabrous, yellowish to reddish at maturity.

UGANDA. Karamoja District: Loyoro, 4 Nov. 1939, *A.S. Thomas* 3157! & Moruangaberu [Emorruangaberru], 9 Feb. 1956, *Dyson-Hudson* 410!; Teso District: Serere, Oct. 1932, *Chandler* 953!
KENYA. Northern Frontier Province: Moyale, 17 Apr. 1952, *Gillett* 12831!; Masai District: R. Uaso Nyiro, 32 km. W. of Magadi, 29 July 1956, *Verdcourt* 1532!; Lamu District: Mambasasa, 17 Oct. 1957, *Greenway & Rawlins* 9362!
TANZANIA. Mbulu District: Lake Manyara National Park, Bagoyo R., 20 Mar. 1964, *Greenway & Kanuri* 11399!; Tanga District: Pongwe–Ngomeni, 13 Jan. 1937, *Greenway* 4850!; Kilosa District: Ruaha R. 2 km. S. of junction with Yovi R., 15 July 1970, *Thulin & Mhoro* 408!; Zanzibar I., Josani Forest, July 1972, *Robins* 58!
DISTR. **U** 1–4; **K** 1–7; **T** 1–8; **Z**; **P**; extending to Egypt, Syria, the Arabian Peninsula, the Cape Verde Is., South Africa and Namibia, also Madagascar and Comoro Is.
HAB. Forest edges, lakesides, riverine, extending into drier country especially where seasonal water collects at foot of hills and scarps, rock outcrops; 0–2200 m.

SYN. *Sycomorus gnaphalocarpa* Miq. in Lond. Journ. Bot. 7: 113 (1848). Type: Ethiopia, Tacazze R., below Djeladjeranne, *Schimper* 874 (L, holo.!, K, P, iso.!)

FIG. 17. *FICUS CAPREIFOLIA* — **1**, leafy twig with figs; **2**, leaf; **3**, **4**, seed flowers; **5**, gall flowers; **6**, staminate flower; **7**, ostiole. 1, 2, 5, 6, from *Aweke* 696; 3, 4, 7, from *W. de Wilde et al.* 7857. Drawn by F.M. Bata-Gillot and I. Zewald.

Ficus gnaphalocarpa (Miq.) A. Rich., Tent. Fl. Abyss. 2: 270 (1851); Peter, F.D.O.-A. 2: 94 (1932); Lebrun & Boutique in F.C.B. 1: 7, photo. 8 (1948); T.T.C.L.: 354 (1949); I.T.U., ed. 2: 250, fig. 55b (1952); F.P.S. 2: 265 (1952); F.W.T.A., ed. 2, 1: 606 (1958); K.T.S.: 317 (1961); Aweke in Meded. Landb. Wageningen 79-3: 29, fig. 7 (1979)
[*F. mucuso* sensu K.T.S.: 318 (1961), *non* Ficalho]
F. sycomorus L. subsp. *gnaphalocarpa* (Miq.) C.C. Berg in Adansonia, sér 2, 20: 272 (1980); C.C. Berg et al. in Fl. Cameroun 28: 130, t. 42 (1985)

NOTE. The form with figs confined to the leafy twigs, found most commonly in the peripheral part of the range, can be recognised as a distinct subspecies: *F. sycomorus* L. subsp. *gnaphalocarpa* (Miq.) C.C. Berg.

5. F. mucuso *Ficalho*, Pl. Ut. Afr. Port.: 27 (1884); Hutch. in F.T.A. 6(2): 98 (1916); Peter, F.D.O.-A. 2: 92 (1932); Lebrun & Boutique in F.C.B. 1: 114 (1948); T.T.C.L.: 355 (1949); I.T.U., ed. 2: 253, photo. 42 (1952); F.W.T.A., ed. 2, 1: 606 (1958); Hamilton, Ug. For. Trees: 98 (1981); C.C. Berg et al. in Fl. Cameroun 28: 132, t. 43 (1985). Type: Angola, Golungo Alto, *Welwitsch* 6416 (K, P, isolecto.!)

Tree up to 30(–40) m. tall, with tall trunk, often with buttresses, main branches ascending. Leafy twigs 3–8 mm. thick, densely minutely white puberulous and with much longer white to dark brown hairs, especially on the nodes, periderm flaking off when dry. Lamina chartaceous to subcoriaceous, subcircular to subcordiform, elliptic or sometimes obovate, 6–17 × 4–15 cm., apex shortly acuminate to acute, base cordate, sometimes truncate, margin entire or crenate, upper surface scabridulous to scabrous, sparsely brownish hirtellous to hirsute on the main veins, lower surface puberulous to hirtellous, on the main veins often densely puberulous and with long white to dark brown hairs with an echinate base; lateral veins 3–6 pairs; petiole (1–)2–9 cm. long, 1–2 mm. thick, densely minutely white puberulous and with much longer white to dark brown hairs with an echinate base, epiderm flaking off when dry; stipules 1–2 cm. long, puberulous to subsericeous, caducous. Figs on branched leafless branchlets up to 30 cm. long on the main and lesser branches or also on the trunk; peduncle 1–2.5 cm. long, 1–2 mm. thick; basal bracts 3–5 mm. long. Receptacle ± depressed globose to obovoid, 2.5–4 cm. in diameter when fresh, 2–3 cm. when dry, ± densely puberulous, red to dark orange at maturity.

UGANDA. Bunyoro District: Budongo Forest, May 1935, *Eggeling* 2029! & Kabwoya, Mar. 1940, *Eggeling* 3860!; Masaka District: Buddu, *Dawe* 309!
TANZANIA. Lushoto District: Korogwe, 26 Aug. 1980, *Archbold* 2783!
DISTR. **U** 2–4; **T** 3; extending to Guinea-Bissau and Angola
HAB. Rain-forest, sometimes persisting in cleared areas; 300–1200 m.

NOTE. Occasionally planted. Whether *Grote* in *Herb. Amani* 4075!, from Tanzania, Amani, without further detail, is native or cultivated is not clear — the same applies to *Archbold* 2783, cited above. The species is probably more common than the number of collections suggests.

6. F. sur *Forssk.*, Fl. Aegypt.-Arab.: cxxiv, 180 (1775); Hutch. in F.T.A. 6(2): 100 (1916); Aweke in Meded. Landb. Wageningen 79-3: 66, fig. 17 (1979); Troupin, Fl. Pl. Lign. Rwanda: 448, fig. 148.6 (1982); C.C. Berg et al. in Fl. Cameroun 28: 135, t. 44 (1985). Type: Yemen, Jiblah [Djöbla], *Forsskål* (C, holo.!)

Tree up to 25(–30) m. tall, sometimes with buttresses. Leafy twigs white to yellowish (or brownish) puberulous to hirtellous or tomentose, partly hirsute or almost glabrous, periderm mostly not flaking off when dry. Lamina chartaceous to coriaceous, elliptic to ovate, subovate or oblong, sometimes suborbicular or lanceolate, 4–20(–32) × 3–13(–16) cm., apex acuminate to acute, base subacute to cordate, margin coarsely crenate-dentate to repand or ± entire; upper surface smooth, sometimes scabrous, glabrous or puberulous on the proximal parts of the main veins, lower surface on the whole surface or only on the main veins puberulous to tomentose or glabrous; lateral veins (3–)5–9 pairs; petiole 1.5–9 cm. long, 1–2 mm. thick, puberulous, hirtellous, hirsute, subtomentose or glabrous, periderm usually not flaking off when dry; stipules 1–3.5 cm. long, white to yellow subsericeous, pubescent, hirsute or almost glabrous, caducous or occasionally subpersistent. Figs on branched leafless branchlets up to 50(–150) cm. long on the older wood, down to the trunk, or occasionally in the leaf-axils; peduncle 0.5–2 cm. long, 1–3 mm. thick; basal bracts 2–3 mm. long; Receptacle obovoid to subglobose, often ± depressed-globose, often stipitate at least when dry, 2–4 cm. in diameter when fresh, 0.5–3 cm. when dry, sparsely white to yellowish puberulous to almost glabrous or densely tomentose to subvelutinous, red to dark orange at maturity. Fig. 18.

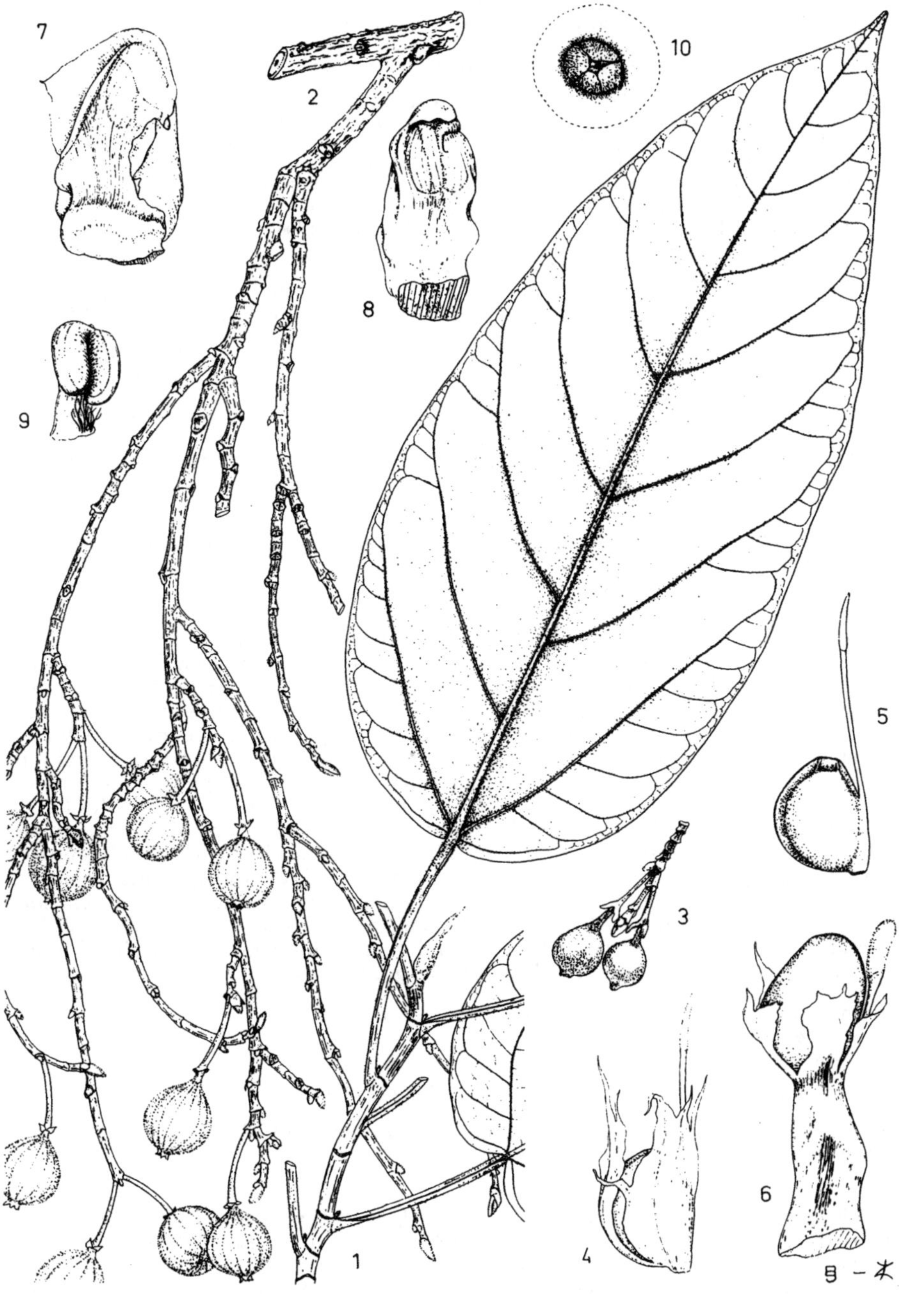

FIG. 18. *FICUS SUR* — **1**, twig with leaves; **2**, fig-bearing branchlet; **3**, young figs; **4**, seed flower; **5**, pistil; **6**, gall flower; **7**, staminate flower, enveloped by bracts; **8**, staminate flower (with saccate perianth); **9**, stamen; **10**, ostiole. 1, 2, from *W. de Wilde et al.* 7206; 3, 10, from *W. de Wilde et al.* 6014; 3–9, from *Aweke & Gilbert* 626.

UGANDA. W. Nile District: Paida, Mar. 1935, *Eggeling* 1919!; Bunyoro District: Budongo Forest, *Eggeling* 3057!; Mengo District: 8 km. on Kampala–Bombo road, Mar. 1938, *Chandler* 2188!
KENYA. Trans-Nzoia District: Kitale, May 1965, *Tweedie* 3047!; Meru District: Nyambeni Hills, Kirima Peak, 9 Oct. 1960, *Verdcourt & Polhill* 2943!; Kiambu District: Karura Forest, 25 Nov. 1966, *Perdue & Kibuwa* 8098!
TANZANIA. Lushoto, 30 June 1966, *Semsei* 4047!; Kondoa District: Kolo, 12 Jan. 1962, *Polhill & Paulo* 1141!; Songea District: R. Luhira, near Mshangano fish ponds, 21 Mar. 1956, *Milne-Redhead & Taylor* 9313!; Zanzibar I., Mazizini [Massazini], 16 May 1959, *Faulkner* 2264!
DISTR. **U** 1–4; **K** 1, 3–7; **T** 1–8; **Z**; **P**; extending to Yemen, Cape Verde Is., Angola and South Africa
HAB. Forest, riverine, wooded grassland, often left in cleared places; 0–2300 m.

SYN. *F. capensis* Thunb., Diss. Fic.: 13 (1786); Hutch. in F.T.A. 6(2): 101 (1916); Peter, F.D.O.-A. 2: 92 (1932); T.T.C.L.: 354 (1949); U.O.P.Z.: 262 (1949); I.T.U., ed. 2: 243, fig. 53 (1952); F.P.S. 2: 265 (1952); F.W.T.A., ed. 2, 1: 606 (1958); K.T.S.: 315 (1961); F.F.N.R.: 30, t. 6C (1962); Hamilton, Ug. For. Trees: 99 (1981). Type: South Africa (not found in UPS-THUNB)
Sycomorus capensis (Thunb.) Miq. in Lond. Journ. Bot. 7: 113 (1848)
S. sur (Forssk.) Miq. in Verh. Eerste Kl. Kon. Ned. Inst. Wet., ser 3, 1: 121 (1849)
Ficus capensis Thunb. var. *trichoneura* Warb. in E.J. 20: 153 (1894). Type: Tanzania, Zanzibar I., *Stuhlmann* 797 (B, holo.!)
F. mallatocarpa Warb. in E.J. 20: 154 (1894): P.O.A. C: 161, t. 9 (1895); Hutch. in F.T.A. 6(2): 97 (1916); Peter, F.D.O.-A. 2: 93 (1932); T.T.C.L.: 355 (1949); Aweke in Meded. Landb. Wageningen 79-3: 41, fig. 10 (1979). Type: Tanzania, Pare District, Ugweno Mts., *Volkens* 465 (B, holo.!, BR, iso.!)
F. plateiocarpa Warb. in E.J. 30: 292 (1901). Type: Tanzania, Uluguru Mts., *Goetze* 2100 (B, holo.!)
F. capensis Thunb. var. *mallotocarpa* (Warb.) Mildbr. & Burret in E.J. 46: 199 (1911)

NOTE. Three more or less deviating forms occur (especially in northern East Africa): a) with the lamina subovate to oblong or lanceolate and its margin entire or nearly so; b) with densely tomentose to subvelutinous figs, and c) with the lamina densely tomentose beneath. These forms have been recognised as distinct taxa.

7. F. vogeliana (*Miq.*) *Miq.* in Ann. Mus. Lugd.-Bat. 3: 295 (1867); Hutch. in F.T.A. 6(2): 94 (1916); F.T.W.A., ed. 2, 1: 606 (1958); C.C. Berg et al. in Fl. Cameroun 28: 140, t. 46 (1985). Type: Fernando Po, *Vogel* 179 (U, lecto.!, K, isolecto.!)

Tree up to 20 m. tall, with buttresses. Leafy twigs 3–5 mm. thick, minutely white puberulous and with much longer yellowish to white hairs; periderm flaking off when dry. Lamina broadly elliptic to obovate or oblong, 5–22 × 2.5–11 cm., apex acuminate, base cordate to subtruncate, margin coarsely dentate to subentire; upper surface scabrous to scabridulous, sparsely, but on the veins more densely, white to yellow hirtellous or hirsute, lower surface white to yellow hirsute, often only on the main veins; lateral veins 5–9 pairs; petiole 0.5–5.5 cm. long, 1–3 mm. thick, minutely white puberulous and with much longer white to yellowish hairs, with epiderm flaking off when dry; stipules 1–2 cm. long, minutely brownish to white puberulous, often also with much longer appressed hairs or sometimes glabrous, subpersistent. Figs on branched leafless branchlets up to 10 m. long arising from the base of the trunk (and then the figs in the litter underneath the tree) or on usually short branchlets up to the main branches; peduncle 0.5–1.5 cm. long, 1–2 mm. thick; basal bracts 3–5 mm. long. Receptacle mostly depressed globose, 2–3 cm. diameter when fresh, 1–2 cm. when dry, minutely puberulous to hirtellous, red to orange-red when mature, often with yellowish to white spots.

UGANDA. Mengo District: Kijude, Dec. 1915, *Dummer* 2733!
DISTR. **U** 4; west to Guinée and Angola
HAB. Swamp forest; 1200 m.

SYN. *Sycomorus vogeliana* Miq. in Lond. Journ. Bot. 7: 112 (1848)

NOTE. This species is recorded from **U** 2, Bwamba, *Lumsden* 5, in I.T.U., ed. 2: 262 (1952) and in Hamilton, Ug. For. Trees: 100 (1981), but specimen not seen.

8. F. vallis-choudae *Del.* in Ann. Sci. Nat., sér. 2, 20: 94 (1843); Hutch. in F.T.A. 6(2): 103 (1916); Peter, F.D.O.-A. 2: 97 (1932); Lebrun & Boutique in F.C.B. 1: 119 (1948); T.T.C.L.: 355 (1949); I.T.U., ed. 2: 261, fig. 57d (1952); F.P.S. 2: 265 (1952); F.W.T.A., ed. 2, 1: 606 (1958); K.T.S.: 323, fig. 61 (1961); F.F.N.R.: 30 (1962); Aweke in Meded. Landb. Wageningen 79-3: 84, fig. 20 (1979); Troupin, Fl. Pl. Lign. Rwanda: 450, fig. 148.7 (1982); C.C. Berg et al. in Fl. Cameroun 28: 142, t. 47 (1985). Type: Ethiopia, Chouda valley, *Galinier* 167 (MPU, holo.!)

Tree up to 15(–20) m. tall. Leafy twigs 2–10 mm. thick, glabrous, sparsely white appressed puberulous or sometimes white hirtellous to tomentose, with periderm flaking off when dry. Lamina coriaceous to subcoriaceous, ovate to cordiform or ± deltoid, 4–24(–44) × 3–24(–30) cm., apex acute to subobtuse or very shortly acuminate, base obtuse to truncate or cordate, margin coarsely and obtusely dentate to repand, sometimes subentire; upper surface smooth and glabrous or puberulous on the main veins, sometimes scabridulous and hirtellous to hispidulous, lower surface glabrous or puberulous, sometimes hirtellous to pubescent; lateral veins 5–8 pairs; petiole 2–11(–13.5) cm. long, 1–3 mm. thick, glabrous, white appressed puberulous or sometimes white hirtellous to tomentose; periderm flaking off when dry; stipules 1–3 cm. long, in the lower part ciliolate, appressed puberulous or subsericeous, caducous. Figs solitary in the leaf-axils or just below the leaves; peduncle 0.2–1.2 cm. long, 4–6 mm. thick; basal bracts ± 0.2 mm. long. Receptacle subglobose to obovoid, 3–6(–10) cm. in diameter when fresh, 1–5 cm. when dry, ± densely white to yellowish puberulous to hirtellous or tomentose, only tomentellous near the ostiole or glabrous, yellowish to orange at maturity with longitudinal orange to reddish stripes.

UGANDA. Acholi District: Agoro, Mar. 1935, *Eggeling* 1729!; Toro District: Kirimia, 21 Nov. 1935, *A.S. Thomas* 1477!; Mengo District: Old Entebbe, Nov. 1931, *Eggeling* 121!
KENYA. Uasin Gishu District: Kipkarren, Dec. 1934, *Dale* in *F.D.* 3404!; Kiambu District: Thika, Athi ridge, Oct. 1948, *H.M. Gardner* 3 in *Bally* 6508!; N. Kavirondo District: Kakamega Forest, Ikuywa R., 6 Jan. 1968, *Perdue & Kibuwa* 9470!
TANZANIA. Mbulu District: Manyara National Park, Main Gate, 21 Jan. 1965, *Greenway & Kanuri* 12059!; Lushoto District: Soni–Mombo, 24 June 1953, *Drummond & Hemsley* 2988!; Ufipa District: Milepa, 13 Oct. 1950, *Bullock* 3445!
DISTR. **U** 1, 2, 4; **K** 1–7; **T** 2–7; extending to Guinée and Mali, Ethiopia, N. Zambia, Malawi, Mozambique and Zimbabwe
HAB. Riverine, lakesides, ground-water forest; 450–1800 m.

SYN. *F. vallis-choudae* Del. var *pubescens* Peter, F.D.O.-A. 2: 98 (1932); T.T.C.L.: 356 (1949). Types: Tanzania, Lushoto District, Sigi Falls, *Peter* 259, Mashewa, *Peter* 13708, between Bungu and Ambangulu, *Peter* 15592 & between Garaya [Ngaraya] and Kwashemshi, *Peter* 15891 (syn. – not found – identity uncertain)

9. F. variifolia *Warb.* in Ann. Mus. Congo., Bot., sér. 6, 1: 30, t. 15 (1904); Hutch. in F.T.A. 6(2): 113 (1916); Lebrun & Boutique in F.C.B. 1: 125 (1948); I.T.U., ed. 2: 261 (1952); F.W.T.A., ed. 2, 1: 606 (1958); C.C. Berg et al. in Fl. Cameroun 28: 145 (1985). Type: Zaire, Monbuttu, Kibali R., *Schweinfurth* 3614 (B, holo.!)

Tree up to 35 m. or more tall, with buttresses. Leafy twigs 2–3 mm. thick, white puberulous or glabrous, when juvenile white hirsute to pubescent. Lamina subcoriaceous, when juvenile chartaceous, oblong to elliptic or ± ovate, 5–20 × 2–11.5 cm., often larger when juvenile, apex subacute to acuminate, base cordate to rounded, margin entire or almost so, when juvenile irregularly lobed to divided; both surfaces glabrous to sparsely puberulous, smooth or scabridulous, when juvenile white hirsute to hirtellous, often scabrous; lateral veins (5–)8–13 pairs or often more when young; petiole 1–5 cm. long, 1–2 mm. thick; stipules 0.3–0.7 cm. long, puberulous or partly hirsute, caducous. Figs solitary or in pairs in the leaf axils; peduncle 0.3–1 cm. long; basal bracts ± 2 mm. long, persistent. Receptacle globose to obovoid, ± 2 cm. in diameter when fresh, ± 1.5 cm. when dry, puberulous, yellow at maturity. Wall of fruiting fig 1.5 mm. thick when dry, firm.

UGANDA. Bunyoro District: Budongo Forest, Nov. 1939, *Eggeling* 3830! & Kitoba Rest Camp, Mar. 1940, *Eggeling* 3859!; Toro District: Kirimia, 21 Nov. 1935, *A.S. Thomas* 1476!
TANZANIA. Kigoma District: Gombe Stream Reserve, Kahama valley, 27 June 1969, *Clutton-Brock* 163!
DISTR. **U** 2, 4; **T** 4; extending to S. Sudan, Angola and Cameroun; also in W. Africa (e.g. in Ivory Coast and Sierra Leone)
HAB. Forest, pioneer of cleared areas; 900–1200 m.

SYN. *F. sciarophylla* Warb. in Ann. Mus. Congo, Bot., sér. 6, 31, t. 13 (1904); F.P.S. 2: 269 (1952). Type: Sudan, Niamniam, Huuh R., *Schweinfurth* 3872 (P, iso.!)

NOTE. The herbarium material from Uganda and Tanzania is for the greater part juvenile.

10. F. dicranostyla *Mildbr.* in E.J. 46: 204 (1911); Hutch. in F.T.A. 6(2): 119 (1916); I.T.U., ed. 2: 247, fig. 54c (1952); F.P.S. 2: 265 (1952); F.W.T.A., ed. 2, 1: 607 (1958); F.F.N.R.: 30 (1962); Aweke in Meded. Landb. Wageningen 79-3: 18, fig. 4 (1979); C.C. Berg et al. in Fl. Cameroun 28: 146, t. 48 (1985). Type: Guinée, Kankan, *Chevalier* 582 (B, holo., not found, P, lecto.!)

Tree up to 10(–20) m. or a shrub. Leafy twigs 2–5 mm. thick, white pubescent to puberulous. Lamina subcoriaceous to chartaceous, elliptic to oblong or ovate, (2–)5–20 × (1–)2–9 cm., apex acuminate, sometimes subacute, base obtuse to acute or ± cordate, margin entire or almost so; upper surface scabrous, hispidulous, on the main veins puberulous to hirtellous or pubescent, lower surface puberulous to hirtellous or pubescent; lateral veins 5–8(–9) pairs; petiole 1–3.5(–5.5) cm. long, 1–2 mm. thick; stipules 0.5–1.5 cm. long, puberulous to pubescent, caducous. Figs solitary or in pairs in the leaf-axils; peduncle 0.3–1 cm. long; basal bracts 2–2.5 mm. long, persistent. Receptacle often shortly stipitate at least when dry, globose to obovoid, 1–2.5 cm. in diameter when fresh, 0.5–1.5 cm. when dry, puberulous to hispidulous, yellowish to pale orange at maturity. Wall of the fruiting fig ± 1.5 mm. thick when dry, firm. Fig. 19.

UGANDA. Acholi District: Agoro, Mar. 1935, *Eggeling* 1714! & Parabong, Apr. 1943, *Purseglove* 1532!; Karamoja, *Eggeling* 2333!

DISTR. U 1–3; extending to S. Ethiopia, S. Sudan, NE. Zaire and Senegal, also in SE. Zaire (Shaba) and N. Zambia

HAB. Wooded grassland, often in rocky places; 900–1100 m.

NOTE. Juvenile specimens, with irregularly lobed leaves, can be confused with *F. asperifolia*. *F. dicranostyla* might prove to be distinct from *F. variifolia* at not more than subspecific level.

11. F. ingens (*Miq.*) *Miq.* in Ann. Mus. Lugd.-Bat. 3: 288 (1867); Hutch. in F.T.A. 6(2): 121 (1916); Peter in F.D.O.-A. 2: 98 (1932); Lebrun & Boutique in F.C.B. 1: 121 (1948); T.T.C.L.: 355 (1949); I.T.U., ed. 2: 250, fig. 56b (1952); F.P.S. 2: 268 (1952); F.W.T.A., ed. 2, 1: 607 (1958); K.T.S.: 317 (1961); F.F.N.R.: 30, fig. 7 (1962); Hamilton, Ug. For. Trees: 100 (1981); Troupin, Fl. Pl. Lign. Rwanda: 444 (1982); C.C. Berg et al. in Fl. Cameroun 28: 149, t. 49 (1985). Type: Ethiopia, Djeladjeranne, *Schimper* 1771 (L, holo.!, B, BR, K, P, iso.!)

Tree up to 18 m. tall. Leafy twigs 3–6 mm. thick, sparsely white or brownish pubescent to densely tomentellous or subvelutinous. Lamina coriaceous, ovate to elliptic or oblong, (2.5–)5–18 × (2–)3–9 cm., apex acuminate to acute, sometimes obtuse, base cordate, occasionally truncate to obtuse, margin entire; both surfaces glabrous; lateral veins 8–11 pairs, the main basal pairs ± distinctly branched, almost straight, thus not running parallel to the margin, the others often branched far from the margin; petiole 0.5–4 cm. long, 1–3 mm. thick; stipules 0.5–1 cm. long, densely yellowish tomentose to subvelutinous or glabrous, caducous. Figs in pairs in the leaf-axils or just below the leaves, subsessile or on peduncles up to 0.5 cm. long; basal bracts ± 2 mm. long. Receptacle ± globose, 1–2 cm. in diameter when fresh, 0.5–1 cm. when dry, minutely white to brown puberulous or partly hirtellous, sometimes densely tomentose or pubescent, whitish to pink or pale to dark purple at maturity; wall wrinked when dry. Fig. 20, p. 62.

UGANDA. Acholi District: Paimol, *Eggeling* 2350!; Teso District: Serere, Sept.–Oct. 1932, *Chandler* 952!; Mbale District: Budama, Mulanda, Sept. 1939, *Dale* U.36!

KENYA. Naivasha District: Mt. Longonot, 5 Dec. 1959, *Polhill* 125!; S. Kavirondo District: Kisii, Mukeria, Sept. 1933, *Napier* in *C.M.* 5274!; Masai District: NW. of Lake Magadi, 28 Dec. 1958, *Greenway* 9551!

TANZANIA. Arusha District: Engare Nanyuki R., 11 Apr. 1968, *Greenway & Kanuri* 13454!; Kondoa District: Bukulu, 14 Jan. 1962, *Polhill & Paulo* 1171!; Morogoro District: above Morogoro, Kibwe rentrant, 23 Sept. 1949, *Wigg* in *F.H.* 3015!; Zanzibar I, Makunduchi, 4 Jan. 1960, *Faulkner* 2453!

DISTR. **U** 1, 3, 4; **K** 1–7; **T** 1–8; **Z**; extending to Yemen, Senegal, Angola, Botswana and South Africa

HAB. Most often on rock outcrops, lava flows, coral and limestone in drier or exposed areas, streamsides, sometimes more extensive stands in disturbed forest, coastal bushland and wooded grassland; 0–2600 m.

SYN. *Urostigma ingens* Miq. in Lond. Journ. Bot. 6: 554 (1847)
Ficus stuhlmannii Warb. var. *glabrifolia* Warb. in E.J. 20: 162 (1894). Type: Tanzania, Mwanza District, Busisi, *Stuhlmann* 750 (B, lecto.!)
F. ingentoides Hutch. in K.B. 1915: 319 (1915); Peter, F.D.O.-A. 2: 99 (1932); T.T.C.L.: 355 (1949); F.P.S. 2: 268 (1952). Type: Ethiopia, Eritrea, near Acrur, *Schweinfurth & Riva* 1687 (K, holo.!)
F. ingens (Miq.) Miq. var. *tomentosa* Hutch. in Fl. Cap. 5(2): 530 (1925); F.W.T.A., ed. 2, 1: 607 (1958); Troupin, Fl. Pl. Lign. Rwanda: 445 (1982). Type: none cited
[*F. lutea* sensu Aweke in Meded. Landb. Wageningen 79-3: 37, fig. 9 (1979), *non* Vahl]

12. F. cordata *Thunb.*, Diss. Fic.: 8, t. (1786); Hutch. in F.T.A. 6(2): 119 (1916) & in Fl. Cap. 5(2): 530 (1925); C.C. Berg et al. in Fl. Cameroun 28: 152, t. 50 (1985). Type: South Africa, *Herb. Thunberg* 24343 (UPS, holo., microfiche!)

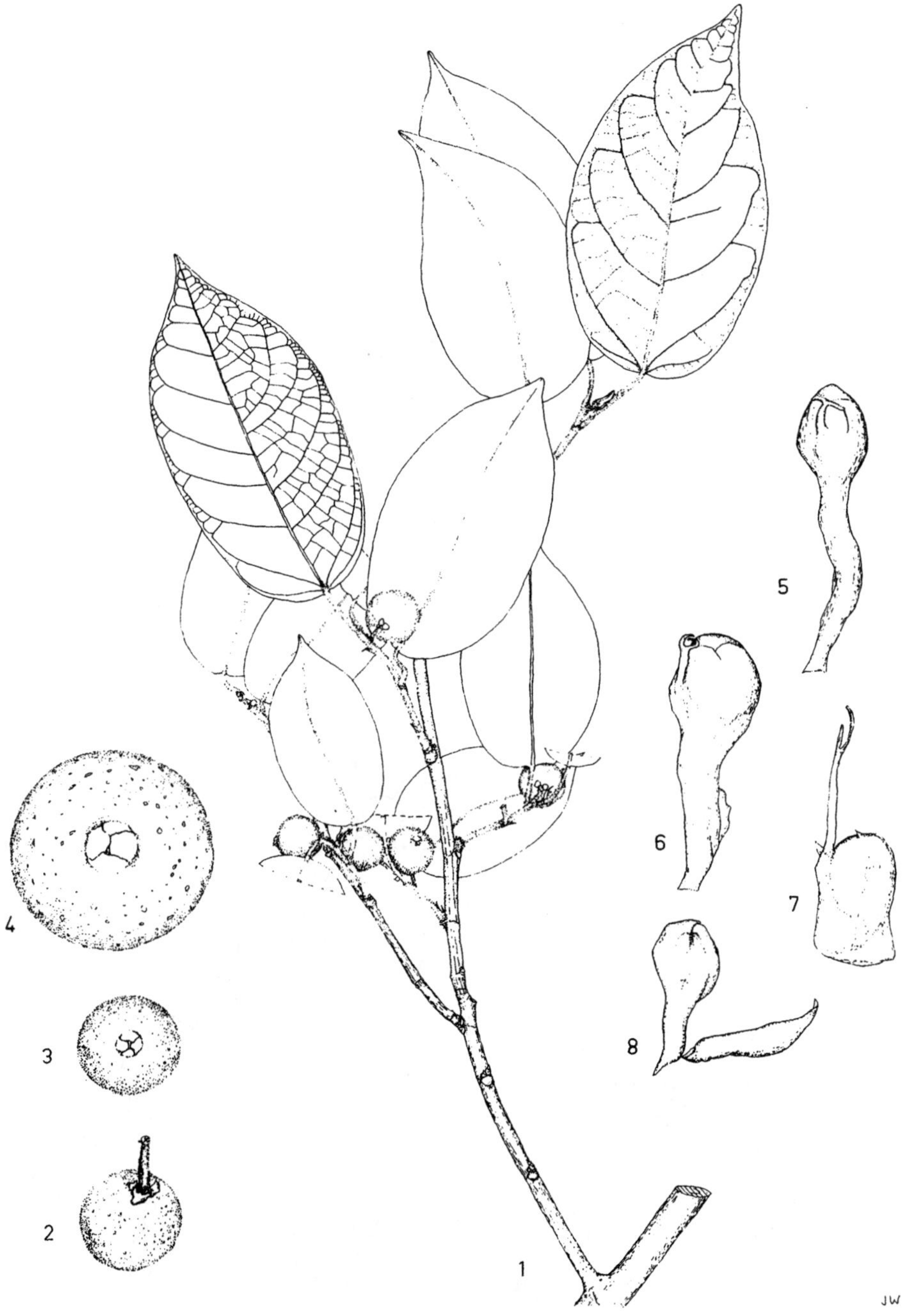

FIG. 19. *FICUS DICRANOSTYLA* — **1**, leafy twig with figs; **2**, fig; **3**, **4**, ostiole; **5**, **6**, gall flower; **7**, seed flower; **8**, staminate flower. All from *Jansen & Aweke* 5087. Drawn by J. Williamson.

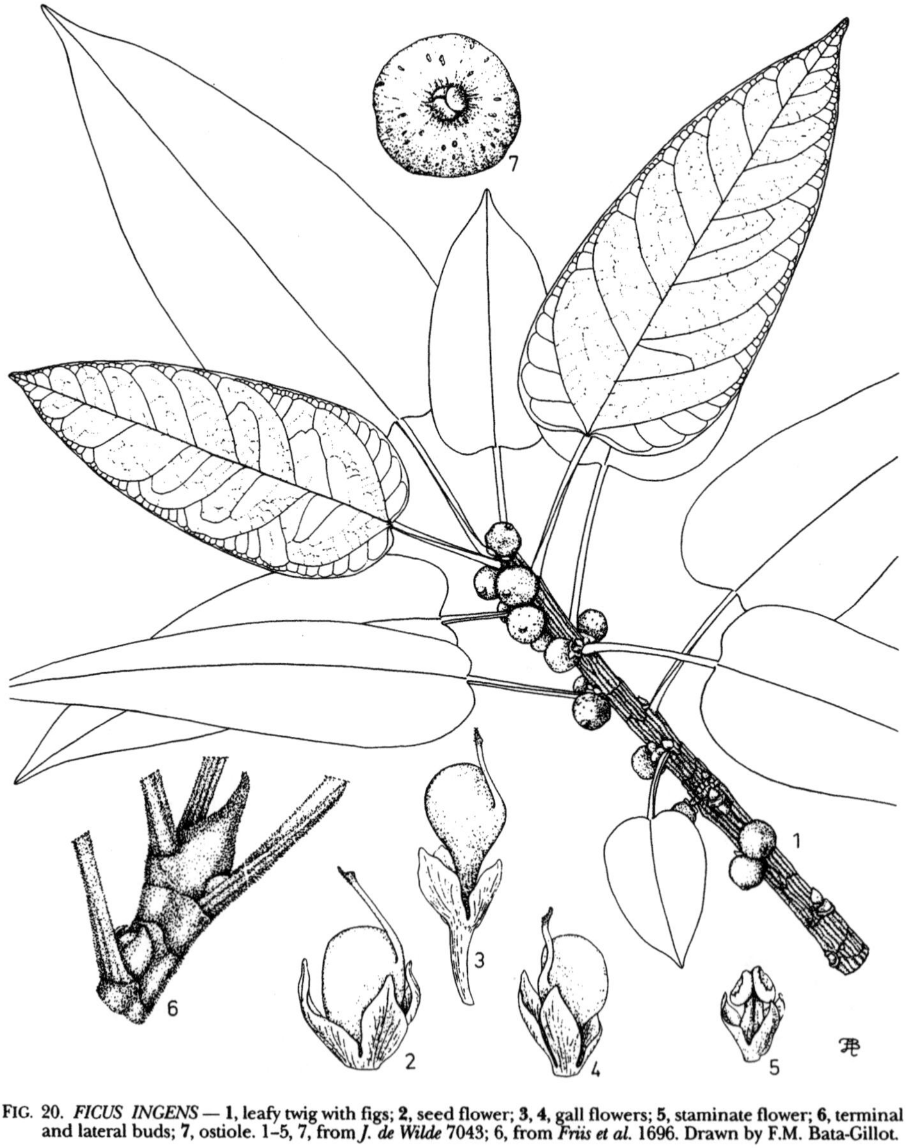

FIG. 20. *FICUS INGENS* — **1**, leafy twig with figs; **2**, seed flower; **3**, **4**, gall flowers; **5**, staminate flower; **6**, terminal and lateral buds; **7**, ostiole. 1–5, 7, from *J. de Wilde* 7043; 6, from *Friis et al.* 1696. Drawn by F.M. Bata-Gillot.

Tree up to 15(–35) m. tall. Leafy twigs 2–5 mm. thick, glabrous to pubescent. Lamina coriaceous, lanceolate, ovate, oblong, elliptic or cordiform, rarely ± obovate, 2–17 × 1–6 cm., apex acuminate, acute, obtuse or occasionally rounded, base cordate to rounded, obtuse or occasionally subacute, margin entire; both surfaces glabrous; lateral veins 6–12 pairs; smaller veins prominent and conspicuous beneath; petiole 0.5–3(–6) cm. long, 1–2 mm. thick; stipules 0.2–1.5 cm. long, glabrous or ciliolate. Figs mostly in the leaf-axils or just below the leaves, or sometimes on older parts up to 3 together on 2–3 mm. long spurs, sessile or on peduncles up to 0.3 cm. long; basal bracts ± 1–1.5 mm. long, sometimes caducous. Receptacle ± globose, 0.5–1 cm. in diameter when fresh, 0.5–0.8(–1) cm. when dry, sparsely and minutely puberulous to glabrous, maturing from green to whitish to dark purple or dark red; wall usually smooth when dry.

subsp. **salicifolia** (*Vahl*) *C.C. Berg* in K.B. 43: 82 (1988). Type: Yemen, *Forsskål* 780 (C, holo.!, B, iso.!)

Leafy twigs usually minutely puberulous or glabrous. Lamina lanceolate to oblong or subovate, sometimes ovate or elliptic; petiole usually drying yellowish to brown. Figs usually with peduncles up to 3 mm. long, sometimes subsessile.

UGANDA. Acholi District: Chua, *Eggeling* 2407!; Karamoja District: Kaabong, 17 Sept. 1950, *Dawkins* 645! & N. of Kotido, Apr. 1960, *J. Wilson* 876!

KENYA. Northern Frontier Province: Isiolo–Archer's Post, Buffalo Springs, 25 Jan. 1961, *Polhill* 334; Nakuru District: Lake Nakuru National Park, 16 Nov. 1973, *Kutilek* 158!; Kiambu District: Ndeiya Grazing Scheme, 20 Jan. 1963, *Verdcourt* 3548!

TANZANIA. Masai District: Engaruka, 9 July 1956, *Bally* 10670! & Olkarien, 20 Dec. 1962, *Newbould* 6398!; Moshi District: Lake Chala, Jan. 1894, *Volkens* 1787!

DISTR. **U** 1; **K** 1, 3, 4, 6, 7; **T** 1, 2; extending to Saudi Arabia, Socotra, North Africa, NE. Zaire, and disjunctly to western Zambia and South Africa (Transvaal)

HAB. Seasonal streams, outcrops of rock, lava or limestone, and local water-catchment areas generally; 950–2400 m.

SYN. *F. salicifolia* Vahl in Symb. Bot. 1: 82, t. 23 (1790); Peter F.D.O.-A. 2: 98 (1932); F.P.S.: 265, fig. 93 (1952)
Urostigma salicifolium (Vahl) Miq. in Lond. Journ. Bot. 6: 556 (1847)
Ficus pretoriae Burtt Davy in Trans. Roy. Soc. S. Afr. 2: 365 (1912); Hutch. in F.T.A. 6(2): 116 (1916); T.T.C.L.: 355 (1949); I.T.U., ed. 2: 257 (1952); F.F.N.R.: 32 (1962); Aweke in Meded. Landb. Wageningen 79-3: 62, fig. 16 (1979). Types: South Africa, Transvaal, Pretoria District, Magaliesberg, *Burtt Davy* 2645 (PRE, syn.), 2750 & 2806 (PRE, syn,!, K, isosyn.!)

NOTE. Subsp. *cordata*, with sessile figs and hairier branchlets, occurs in South Africa (Cape Province), Namibia, NW. Botswana and Angola; subsp. *lecardii* (Warb.) C.C. Berg, with broader leaves and petiole drying darker, occurs in W. Africa from Senegal through northern Cameroun to the Central African Republic.

13. F. verruculosa *Warb.* in E.J. 20: 166 (1894); Hutch. in F.T.A. 6(2): 114 (1916); Peter, F.D.O.-A. 2: 99 (1932); Lebrun & Boutique in F.C.B. 1: 120 (1948); T.T.C.L.: 356 (1949); I.T.U., ed. 2: 262 (1952); F.W.T.A., ed. 2, 1: 607 (1958); K.T.S.: 323 (1961); F.F.N.R.: 32, t. 6E (1962); Troupin, Fl. Pl. Lign. Rwanda: 450, fig. 148.4 (1982); C.C. Berg et al. in Fl. Cameroun 28: 154, t. 51 (1985). Type: Angola, Huila, between Monino and Eime, *Welwitsch* 6375 (B, holo.!, K, P, iso.!)

Shrub or treelet up to 7 m. tall. Leafy twigs 1–5 mm. thick, glabrous or densely white hirtellous to subtomentellous. Lamina coriaceous, oblong to lanceolate, 3.5–10(–20) × 1.5–3.5(–8.5) cm., apex subacute to obtuse, base obtuse or rounded to subcordate, margin entire; both surfaces glabrous; lateral veins (8–)10–16 pairs, the basal pair unbranched, curved and running almost parallel to the margin, the others divided rather near the margin; only the midrib prominent beneath, the other veins often plane and inconspicuous; petiole 0.5–2(–3) cm. long, 1–2 mm. thick; stipules 0.5–3.5(–4) cm. long, glabrous to densely white puberulous, caducous. Figs mostly in pairs in the leaf-axils or just below the leaves; peduncle 0.3–0.5(–1) cm. long, 1–1.5 mm. thick; basal bracts ± 1 mm. long. Receptacle subglobose or sometimes ellipsoid, 0.5–2 cm. in diameter when fresh, 0.5–1(–1.2) cm. when dry, glabrous or minutely puberulous, maturing dark purple or dark red; wall often wrinkled when dry.

UGANDA. Kigezi District: Butali, 13 Aug. 1936, *A.S. Thomas* 2032!; Mbale District: Elgon, Siroko Valley, 21 Feb. 1924, *Snowden* 843!; Masaka District: Lake Nabugabo, June 1937, *Chandler* 1707!

KENYA. Trans-Nzoia District: Kitale, 13 May 1953, *Bogdan* 3733!

TANZANIA. Kigoma District: Kitolo, 8 Oct. 1949, *Bally* 7632!; Rungwe District: Mulinda, 4 Nov. 1912, *Stolz* 1639!; Songea District: Nonganonga stream, 28 Dec. 1955, *Milne-Redhead & Taylor* 7946!

DISTR. **U** 2–4; **K** 3; **T** 4, 7, 8; extending to Niger, Angola and South Africa

HAB. Streams, lakes, swamps, bogs, commonly in water, occasionally recorded from grassland or wooded grassland; 800–1850 m.

14. F. platyphylla *Del.*, Cent. Pl. Méroé: 62 (1826); Hutch. in F.T.A. 6(2): 198 (1917); I.T.U., ed. 2 : 256 (1952); F.P.S. 2: 272 (1952); F.W.T.A., ed. 2, 1: 609 (1958); Aweke in Meded. Landb. Wageningen 79-3: 54, fig. 13 (1979); Friis in Nordic Journ. Bot. 5: 333 (1985); C.C. Berg et al. in Fl. Cameroun 28: 165, t. 55 (1985). Type: Sudan, Méroé, *Cailliaud* (MPU, holo.!)

Tree up to 15 m. tall, (secondarily ?) terrestrial. Leafy twigs (5–)10–20 mm. thick, densely white puberulous or velutinous, on the nodes pubescent; periderm flaking off when dry. Leaves in spirals; lamina coriaceous, ovate to elliptic or subovate to oblong, (8–)15–26 × (5–)10–20 cm., apex obtuse to acute, base cordate; margin entire to repand; upper surface glabrous or puberulous on the lower part of the midrib, lower surface white puberulous to pubescent or only puberulous on the main veins; lateral veins 10–16 pairs, the basal pair branched, reaching the margin far below the middle of the lamina; tertiary venation partly scalariform; petiole 4–10 cm. long, 2–5 mm. thick, periderm not flaking off when dry; stipules 0.8–3.5 cm. long, densely white tomentose to puberulous to subvelutinous, caducous. Figs up to 5 together in the leaf-axils or just below the leaves; peduncle 1–2.5 cm. long; basal bracts ± 3 mm. long, persistent. Receptacle globose, 1–2 cm. in diameter when fresh, ± 1(–1.5) cm. when dry, puberulous, warted, greenish at maturity.

UGANDA. Acholi District: Padibe, Apr. 1938, *Eggeling* 3576; Karamoja District, *Eggeling* 2334!; Teso District: Serere, Sept. 1932, *Chandler* 883!
DISTR. **U** 1, 3; extending to Somalia, Ethiopia and Senegal
HAB. Wooded grassland, rocky places; 950–1170 m.

15. F. bussei *Mildbr. & Burret* in E.J. 46: 213 (1911); Peter, F.D.O.-A. 2: 102 (1932); T.T.C.L.: 360 (1949); K.T.S.: 315 (1961); Friis in Nordic Journ. Bot. 5: 332 (1985). Type: Tanzania, Lindi District, road to Kitulu, *Busse* 2427 (B, lecto.!, BR!, EA, isolecto.)

Tree up to 25 m. high, (secondarily ?) terrestrial. Leafy twigs 4–12 mm. thick, sparsely to sometimes densely puberulous to hirtellous or glabrous, periderm hardly flaking off when dry. Leaves in spirals; lamina coriaceous (often brittle when dry), subovate to oblong, 5–24 × 3–9.5(–11.5) cm., apex acute to obtuse, base cordate, margin entire to repand; upper surface glabrous or puberulous on the midrib, lower surface almost glabrous or ± sparsely hirtellous to puberulous; lateral veins 10–16 pairs, the basal pairs reaching the margin far below the middle of the lamina; tertiary venation partly scalariform; petiole 2–8 cm. long, 2–4 mm. thick, epiderm not flaking off; stipules 0.3–1.2(–5) cm. long, glabrous or pubescent at the base, caducous. Figs in pairs or solitary in the leaf-axils; peduncle 1–2.5 cm. long, recurved; basal bracts ± 3 mm. long, persistent. Receptacle ± globose or rarely ellipsoid, 2–3 cm. in diameter when fresh, 1–1.5 cm. when dry, puberulous, smooth or warted, ? greenish at maturity.

KENYA. Teita District: Ndara Plain on Voi R. crossing, 8 Dec. 1966, *Greenway & Kanuri* 12703!; Mombasa, N. Harbour, 24 Oct. 1977, *Gillett & Stearn* 21630!; Kilifi District: Kiboriani, 20 Mar. 1946, *Jeffery* 497!
TANZANIA. Lushoto District: Mkuye, 2 Feb. 1941, *Greenway* 6011! & Korogwe, 6 Aug. 1970, *Archbold* 2672!; Mpwapwa District: Matomondo [Matamanda], 2 Sept. 1937, *Hornby* 840!; Lindi District: Kitulu [Kitulo], 11 May 1903, *Busse* 2472!
DISTR. **K** 7; **T** 3, 5, 6, 8; Somalia, Mozambique, Malawi, Zambia and Zimbabwe
HAB. Lowland forest, riverine, swamp forest and flood plains; 0–550 m.

SYN. *F. fasciculata* Warb. in E.J. 20: 175 (1894), *non* Benth. (1873), *nom. illegit.* Type: Tanzania, Zanzibar, Changa [Changu] I., *Stuhlmann* 109 (B, holo.!)
F. changuensis Mildbr. & Burret in E.J. 46: 212 (1911); Peter, F.D.O.-A. 2: 104 (1932); T.T.C.L.: 360 (1949). Type as for *F. fasciculata*
F. zambesiaca Hutch. in K.B. 1915: 341 (1915) & in F.T.A. 6(2): 198 (1917); F.F.N.R.: 33 (1962). Type: Malawi, Shire valley, Katungu, *Scott* (K, lecto.!)

NOTE. *F. bussei*, the west and central African *F. recurvata* De Wild., and *F. platyphylla* (as well as *F. vasta*) form a group of very closely related species, with rather weak differentiating characters.

16. F. vasta *Forssk.*, Fl. Aegypt.-Arab.: cxxiv, 179 (1775); Hutch. in F.T.A. 6(2): 194 (1917); Peter, F.D.O.-A. 2: 101 (1932); I.T.U., ed. 2: 262 (1952); F.P.S. 2: 271 (1952); K.T.S.: 323 (1961); Aweke in Meded. Landb. Wageningen 79-3: 88, fig. 21 (1979). Type: Yemen, *Forsskål* 776 (C, holo.!)

Tree up to 25 m. tall, (secondarily ?) terrestrial. Leafy twigs 6–15 mm. thick, yellowish to whitish or brownish hirsute to hirtellous, periderm flaking off when dry. Leaves in spirals; lamina coriaceous, cordiform to ovate, subcircular or elliptic to subreniform, (5–)8–25(–35) × (2.5–)4–23(–25) cm., apex rounded to obtuse or very shortly and bluntly acuminate, base cordate, margin entire or almost so; upper surface rather sparsely hirtellous to puberulous or almost glabrous, lower surface whitish subvelutinous or densely to sparsely hirtellous, puberulous or almost glabrous; lateral veins 5–10 pairs, the basal pairs branched, reaching the margin usually at or above the middle of the lamina; tertiary venation for the greater part scalariform; petiole (1.5–)3–12(–19) cm. long, 1.5–4 mm. thick, periderm flaking off when dry; stipules 2–5 cm. long, on flush up to 8.5 cm., whitish to yellowish or brownish subhirsute to subsericeous, caducous. Figs in pairs or solitary in the leaf-axils or just below the leaves, subsessile or on peduncles up to 6 mm. long; basal bracts 3.5–4.5 mm. long, free parts caducous. Receptacle subglobose to ellipsoid, 2–2.5 cm. in diameter when fresh, 1–1.5 cm. when dry, whitish to yellowish velutinous or sparsely hirtellous, often ± verruculate, green with paler spots at maturity.

UGANDA. Acholi District: Agoro, Mar. 1935, *Eggeling* 1727!; Karamoja District: Warr, 4 Nov. 1939, *A.S. Thomas* 3174!
KENYA. Northern Frontier Province: Moyale, 18 Apr. 1952, *Gillett* 12838!
DISTR. **U** 1; **K** 1; Sudan, Ethiopia, Somalia, Saudi Arabia, N. and S. Yemen, Socotra
HAB. Riverine; 1100–1650 m.

17. F. wakefieldii *Hutch.* in K.B. 1915: 335, fig. (1915) & in F.T.A. 6(2): 168 (1916); Peter, F.D.O.-A. 2: 101 (1932); T.T.C.L.: 359 (1949); K.T.S.: 323 (1961); F.F.N.R.: 32 (1962). Type: Kenya, coast, without locality, *Wakefield* 34 (K, lecto.!)

Tree up to 25 m. tall, secondarily ? terrestrial, with a wide crown. Leafy twigs (3–)5–12 mm. thick, with minute to short hairs, intermixed with much longer yellow to brownish hairs, periderm flaking off when dry. Leaves in spirals; lamina ± coriaceous, cordiform to ovate, broadly elliptic, subcircular, subreniform or sometimes broadly obovate, 6–23 × 5–23 cm., apex rounded or sometimes very shortly and bluntly acuminate, base cordate, margin entire or almost so; upper surface sparsely hirtellous to subhirsute, lower surface sparsely to densely hirtellous to puberulous, on the main veins yellow hirsute; lateral veins 5–8 pairs, main basal pair branched, reaching the margin at or just below the middle of the lamina; tertiary venation partly scalariform; petiole 2–5.5(–9) cm. long, (1.5–)2–4(–5) mm. thick, epiderm flaking off; stipules 0.5–1.5 cm. long, on flush up to 4 cm., yellow to brownish hirsute to subsericeous, caducous. Figs in pairs in the leaf-axils, initially enclosed in ovoid calyptrate buds up to 1.5 cm. long, sessile; basal bracts 3–5 mm. long, persistent. Receptacle globose or almost so, (1.2–)1.5–2 cm. in diameter when fresh, (0.8–)1–1.5 cm. when dry, densely white to yellow pubescent to ± hirsute or sparsely hirtellous.

UGANDA. Acholi District: Rom, *Liebenberg* 302! (leaves only); Busoga District: Lolui I., 16 May 1974, *G. Jackson* U.116!
KENYA. Northern Frontier Province: SW. foothills of Ndoto Mts., Ngurunit Mission, 15 Jan. 1979, *Synnott* 1841!; Naivasha District: Lake Naivasha, Mennells's Farm, 12 Feb. 1982, *Mwangangi* 2178!; Kiambu District: Thika, Athi ridge, Oct. 1948, *H.M. Gardner* 7 in *Bally* 6512!
TANZANIA. Mwanza District: 19 km. Mwanza–Shinyanga, 17 July 1960, *Verdcourt* 2892!; Musoma District: Mto ya Mchanga, 17 Feb. 1968, *Greenway et al.* 13301!; Kondoa District: Kolo, 16 Jan. 1928, *B.D. Burtt* 1102!
DISTR. **U** 1, 3; **K** 1, 3, 4, 6, 7; **T** 1, 2, 5, 7; Zambia
HAB. Riverine, lakesides, wooded grassland, commonly on rock outcrops, scarps or at foot of rocky hills; 200–2000 m.

NOTE. Closely related to *F. vasta*.

18. F. glumosa *Del.*, Cent. Pl. Méroé: 63 (1826); Hutch. in F.T.A. 6(2): 171 (1916); Peter, F.D.O.-A. 2: 104 (1932); Lebrun & Boutique in F.C.B. 1: 147 (1948); T.T.C.L.: 359 (1949); I.T.U., ed. 2: 248, fig. 55a (1952); F.P.S. 2: 270 (1952); F.W.T.A., ed. 2, 1: 609 (1958); K.T.S.: 609 (1961); Aweke in Meded. Landb. Wageningen 79-3: 25, fig. 6 (1979); Troupin, Fl. Pl. Lign. Rwanda: 444, fig. 148.5 (1982); C.C. Berg et al. in Fl. Cameroun 28: 170, t. 58 (1985). Type: Ethiopia, without locality, *Cailliaud* (MPU, holo.!)

Tree up to 10(–15) m. tall or a shrub, terrestrial, with spreading branches. Leafy twigs 2–6 mm. thick, with dense short white hairs, especially on the nodes intermixed with much longer yellow to sometimes whitish hairs or glabrous, periderm of older parts flaking off when dry. Leaves in spirals; lamina subcoriaceous, oblong to elliptic or broadly ovate, sometimes obovate or subcircular, 2–14 × 1.2–9.5 cm., apex shortly acuminate to subacute or subobtuse, base cordate, sometimes rounded, margin entire; upper surface puberulous to hirtellous to subtomentose, or almost glabrous, lower surface subtomentose on the main veins, often partly yellow hirsute or entirely glabrous; lateral veins 3–7 pairs, the basal pairs branched, sometimes faintly, reaching the margin below or sometimes at the middle of the lamina, tertiary venation reticulate or tending to scalariform, the smaller vein-reticulum beneath inconspicuous; petiole 0.5–4(–8) cm. long, 1–2(–2.5) mm. thick, epiderm not flaking off; stipules 0.5–1.5(–2.4) cm. long, on flush up to 4 cm., at least partly yellow or sometimes whitish hirsute to subsericeous, caducous. Figs in pairs in the leaf-axils or down to ± 1 m. below the leaves, sessile or sometimes on peduncles up to 3 mm. long; basal bracts ± 3 mm. long, persistent. Receptacle globose or ellipsoid, ± 1–1.5(–2) cm. in diameter when fresh, ± 0.5–1 cm. when dry, densely tomentose to pubescent to almost glabrous, orange to pink or red at maturity, often with darker spots.

UGANDA. W. Nile District: Metuli, 25 Nov. 1941, *A.S. Thomas* 4057!; Acholi District: Abbia Ferry, Mar. 1935, *Eggeling* 1690!; Teso District: Serere, Mar. 1932, *Chandler* 652!

KENYA. Northern Frontier Province: Moyale, 18 Apr. 1952, *Gillett* 12840!; Machakos District: 157 km. Nairobi–Kibwezi, 25 Apr. 1969, *Napper & Kanuri* 2095!; Teita District: Mudanda [Mutanda] Rock, 11 Jan. 1967, *Greenway & Kanuri* 13031!

TANZANIA. Shinyanga District: 13 km. W. of Old Shinyanga, Elephant Rock, 13 Apr. 1958, *Welch* 464!; Maswa District: Moru Kopjes, 11 Apr. 1962, *Greenway et al.* 10596!; Kilosa District: Ilonga, Matarawe R., 25 May 1968, *Renvoize & Abdallah* 2371!

DISTR. **U** 1, 3, 4; **K** 1, 2–7; **T** 1–8; extending to Senegal, Yemen and South Africa

HAB. Rocky outcrops, scarps, lava flows in wooded grassland and deciduous bushland; 500–2000 m.

SYN. *Sycomorus hirsuta* Sond. in Linnaea 23: 137 (1850), *non Ficus hirsuta* Vell. (1825). Type: South Africa, Durban [Port Natal], *Gueinzius* 415 (U, iso.!)

Ficus sonderi Miq. in Ann. Mus. Lugd.-Bat. 3: 295 (1867); Hutch. in F.T.A. 6(2): 170 (1916); Peter, F.D.O.-A. 2: 107 (1932); T.T.C.L.: 359 (1949); I.T.U., ed. 2: 258 (1952); K.T.S.: 320 (1961); F.F.N.R.: 32 (1962). Type as *Sycomorus hirsuta*

F. rukwaensis Warb. in E.J. 30: 295 (1901). Type: Tanzania, Chunya District, Uwungu [Ubungu], Home Mt., *Goetze* 1100 (BM, BR, P, iso.!)

F. glumosa Del. var. *glaberrima* Martelli, Fl. Bogos.: 76 (1886); Hutch. in F.T.A. 6(2); 172 (1916); F.W.T.A., ed. 2, 1: 609 (1958). Type: Ethiopia, Eritrea, Cheren, *Beccari* 40 (?FT, holo., BM, iso.!)

NOTE. In Uganda and in parts of Kenya a form with glabrous or almost glabrous leaves (described as var. *glaberrima* Martelli) is found.

19. F. stuhlmannii *Warb.* in E.J. 20: 161 (1894); P.O.A. C: 162, t. 11F, G (1895); Peter, F.D.O.-A. 2: 107 (1932); T.T.C.L.: 359 (1949); I.T.U., ed. 2: 258 (1952); K.T.S.: 320 (1961); F.F.N.R.: 33 (1962). Type: Tanzania, Shinyanga District, Njangesi, *Stuhlmann* 4141 (B, holo.!)

Tree up to 10(–15) m. tall, terrestrial or hemi-epiphytic (and strangling). Leafy twigs (2–)4–8 mm. thick, rather densely white puberulous to hirtellous, on the nodes to pale yellow hirsute, periderm of older parts ± flaking off when dry. Leaves in spirals; lamina ± coriaceous, oblong or elliptic to ovate or ± obovate, sometimes to subcircular, 2.5–18 × 1–9 cm., apex rounded to subacute or sometimes very shortly and bluntly acuminate, base cordate to rounded, margin ± entire; upper surface puberulous to hirtellous, lower surface densely hirtellous to subtomentose on the veins; lateral veins (3–)4–7 pairs, the basal pair not or faintly branched, usually reaching the margin rather far below the middle of the lamina, tertiary venation reticulate and prominent; petiole 0.5–4(–5.5) cm. long, (1–)2–3 mm. thick, epiderm not flaking off; stipules 0.5–1.5 cm. long, white to pale yellow subsericeous, subhirsute or puberulous, caducous. Figs in pairs in the leaf-axils, sometimes also just below the leaves, sessile or almost so; basal bracts ± 3 mm. long. Receptacle globose to ellipsoid, 1.5–2.2 cm. in diameter when fresh, 0.7–1.8 cm. when dry, densely white pubescent to sparsely puberulous, pinkish or purplish at maturity.

UGANDA. Karamoja District: Kakamari, June 1930, *Liebenberg* 391! & Warr, 4 Nov. 1939, *A.S. Thomas* 3175! & Moroto R., May 1948, *Eggeling* 5788!

KENYA. Kiambu District: Thika, Athi Ridge, 6 Oct. 1948, *H.M. Gardner* 6 in *Bally* 6511!; Kitui District: Mutomo Hill, 20 Jan. 1942, *Bally* 1582!; Mombasa District: Shimo la tewa, Feb. 1937, *Dale* in *F.D.* 3634!

TANZANIA. Ufipa District: Namwele, 30 Dec. 1961, *Richards* 15818!; Kondoa District: Kolo, 12 Jan. 1962, *Polhill & Paulo* 1140!; Iringa, 15 July 1956, *Milne-Redhead & Taylor* 11213!

DISTR. **U** 1; **K** 3, 4, 7; **T** 1–8; Zaire (Shaba), Zambia, Malawi, Mozambique, Zimbabwe, Botswana and South Africa

HAB. Wooded grassland and bushland, often along lakesides, watercourses, on rock or coral outcrops, hardpans; 0–1800 m.

SYN. *F. dar-es-salaamii* Hutch. in F.T.A. 6(2): 171 (1916); Peter, F.D.O.-A. 2: 107 (1932); T.T.C.L.: 358 (1949). Types: Tanzania, Uzaramo District, Dar es Salaam, *Holtz & Stuhlmann* 923 (B, syn.)

20. F. nigro-punctata *Mildbr. & Burret* in E.J. 46: 220, fig. 3 (1911); Hutch. in F.T.A. 6(2): 173 (1916); Peter, F.D.O.-A. 2: 109 (1932); T.T.C.L.: 359 (1949). Type: Tanzania, Lindi District, Seliman-Mamba, road to Kwa-Mbua, *Busse* 2801 (B, lecto.!, EA, isolecto.)

Shrub or tree up to 7 m. tall, terrestrial or sometimes hemi-epiphytic (and strangling). Leafy twigs 1–3(–6) mm. thick, puberulous to hirtellous or subtomentellous, periderm not flaking off, bark of the older wood often blackish and conspicuously lenticellate. Leaves in spirals; lamina chartaceous to subcoriaceous, oblong to elliptic, ± obovate or sometimes ovate, 1–9.5 × 0.6–5.5 cm., apex shortly acuminate to subacute, base rounded to cordate, margin crenulate; upper surface puberulous to hirtellous or hispidulous, when dry sometimes black-punctate, lower surface puberulous to hirtellous on the veins; lateral veins 3–5(–6) pairs, the basal pairs faintly branched, usually reaching the margin ± far below the middle of the lamina, tertiary venation reticulate; petiole 0.3–2 cm. long, 0.5–1 mm. thick, epiderm not flaking off; stipules 0.2–1.2 cm. long, sparsely puberulous to hirtellous, caducous. Figs in pairs in the leaf-axils or also on the older wood, sessile; basal bracts 2–2.5 mm. long, persistent. Receptacle subglobose, 1–1.2 cm. in diameter when fresh, 0.5–1 cm. when dry, puberulous to hirtellous, green with red spots or reddish at maturity.

KENYA. Kitui District: 2 km. Mutomo–Mutha, 22 Nov. 1979, *Gatheri et al.* 79/124!
TANZANIA. Shinyanga, 3 Feb. 1931, *B.D. Burtt* 3287!; Mpwapwa, 27 Jan. 1935, *Hornby* 612!; Rufiji District: Beho Beho, 9 June 1977, *Vollesen* in *MRC*.4649!
DISTR. **K** 4; **T** 1, 3, 5, 6, 8; Zambia, Malawi, Mozambique, Zimbabwe and Botswana
HAB. Woodland, bushland and thicket, often riparian or on rock outcrops; near sea-level to 1300 m.

21. F. abutilifolia (*Miq.*) *Miq.* in Ann. Mus. Lugd.-Bat. 3: 288 (1867); Hutch. in F.T.A. 6(2): 191 (1916); F.P.S. 2: 272 (1952); F.W.T.A., ed. 2, 1: 609 (1958); Aweke in Meded. Landb. Wageningen 79-3: 9, fig. 1 (1979); C.C. Berg et al. in Fl. Cameroun 28: 162, t. 54 (1985). Type: Sudan, Fazogli [Fazokal], *Kotschy* 462 (K, holo.!, B, BM, P, iso.!)

Tree up to 15 m. tall, terrestrial (often epilithic). Leafy twigs 6–10 mm. thick, glabrous or yellowish to white tomentose, tomentellous or puberulous, periderm often flaking off when dry. Leaves in spirals; lamina ± coriaceous, cordiform to broadly ovate or subreniform, 6–17 × 5–18 cm., apex shortly acuminate to subacute, obtuse or rounded, base cordate, margin entire; upper surface glabrous or with sparse hairs on the main veins, lower surface puberulous to subtomentellous, sometimes only in the axils of the lateral veins (or occasionally glabrous); lateral veins 7–9 pairs, the basal pair branched, reaching the margin at or above the middle of the lamina, tertiary venation partly scalariform; petiole 2–10(–18) cm. long, 2–4 mm. thick, epiderm not flaking off; stipules 0.5–2 cm. long, puberulous or glabrous, caducous. Figs in pairs in the leaf-axils or just below the leaves; peduncle 0.3–1.5 cm. long; basal bracts 3–3.5 mm. long, free parts caducous. Receptacle subglobose, obovoid or ellipsoid, 1.2–2 cm. in diameter when fresh, 0.5–1.5 cm. when dry, sparsely minutely puberulous, reddish or yellowish at maturity.

UGANDA. Mengo District: Entebbe, *Dawe* 938!
KENYA. Turkana District: Lodwar, Nyao, 19 Sept. 1963, *Paulo* 959!
TANZANIA. Singida District: Sari Hill, above Lake Kitangiri, 2 Nov. 1960, *Richards* 13478!; Lindi District: Nachingwea, 20 Nov. 1952, *Anderson* 818!
DISTR. **U** 4; **K** 2; **T** 5, 8; extending to Guinée and Ethiopia, also disjunctly to Mozambique, S. Zambia, Zimbabwe, Botswana and South Africa
HAB. Riverine, lakesides, rock outcrops and scarps; 550–1050 m.

SYN. *Urostigma abutilifolium* Miq. in Verh. Eerste Kl. Kon. Ned. Inst. Wet., ser 3, 1: 133, t. 3 (1849)
Ficus discifera Warb. in E.J. 36: 210 (1905); I.T.U., ed. 2: 248 (1952). Type: Sudan, between Gedaref and Abu Harraz, Jebel Arrong, *Schweinfurth* 548 (B, holo.!, P, iso.!)
F. soldanella Warb. in Viert. Nat. Ges. Zürich 51: 136 (1906). Type: South Africa, Transvaal, Pretoria, Kuduspoort, *Rehmann* 4684 (Z, holo.!, B, iso.!)

NOTE. The material from the northern part of the range has caducous basal bracts, whereas that from the southern part usually has persistent ones.

22. F. populifolia *Vahl*, Symb. Bot. 1: 82, t. 22 (1790); Hutch. in F.T.A. 6(2); 189 (1916); Peter, F.D.O.-A. 2: 103 (1932); T.T.C.L.: 361 (1949); U.O.P.Z.: 263 (1949); I.T.U., ed. 2: 257, fig. 55c (1952); F.P.S. 2: 270, fig. 96 (1952); F.W.T.A., ed. 2, 1: 609 (1958); K.T.S.: 319 (1961); Aweke in Meded. Landb. Wageningen 79-3: 56, fig. 14 (1979); C.C. Berg et al. in Fl. Cameroun 28: 160, t. 53 (1985). Type: plate 22 in Symb. Bot. 1 (1790), probably drawn from *Forsskål*, Yemen, Wadi Zebid – type of *F. religiosa* sensu Forssk. (1775) *non* L. (1753)

Tree up to 10(–? 30)m. tall, terrestrial or a shrub. Leafy twigs 3–10 mm. thick, glabrous or white puberulous, periderm not flaking off. Leaves in spirals; lamina ± coriaceous, cordiform to broadly ovate or ± reniform, 3–18 × 3–14.5 cm., apex ± acuminate to sometimes caudate, base cordate or truncate, margin entire to repand; both surfaces glabrous; lateral veins 6–11 pairs, the basal pair branched, reaching the margin below, or sometimes at the middle of the lamina; tertiary venation partly scalariform; petiole 2.5–12 cm. long, 1–2 mm. thick, epiderm not flaking off; stipules 1–4 cm. long, glabrous or ciliolate, caducous. Figs in pairs in the leaf-axils; peduncle 0.8–2 cm. long; basal bracts ± 2 mm. long, free parts caducous. Receptacle obovoid to ellipsoid or subglobose, 1–1.5 cm. in diameter when fresh, 0.5–1 cm. when dry, minutely puberulous to glabrous, often with distinct ribs when dry, green with red spots at maturity.

UGANDA. Acholi District: Agoro, Mar. 1935, *Eggeling* 1726!; Karamoja District: R. Apule, 28 Oct. 1939, *A.S. Thomas* 3089! & Matheniko, Sept. 1943, *Dale* U.359!

KENYA. Northern Frontier Province: Kailongol Mts., 15 June 1970, *Mathew* 6819!; Masai District: 69 km. Nairobi–Magadi, 8 Jan. 1958, *Greenway* 9511!; Teita District: Voi, 9 Dec. 1961, *Polhill & Paulo* 938!

TANZANIA. Mwanza District: Speke Gulf near Mwanza, 31 May 1931, *B.D. Burtt* 2485! & Mbarika, 10 May 1953, *Tanner* 1477!; Arusha District: 24 km. S. of Longido, 29 May 1965, *Leippert* 5779!

DISTR. **U** 1; **K** 1–4, 6, 7; **T** 1, 2; extending to Ghana and Yemen

HAB. Riparian, rock outcrops, lava cliffs and scarps in deciduous bushland and wooded grassland; 500–1600 m.

SYN. *F. populifolia* Vahl var. *taitensis* Warb. in E.J. 36: 212 (1905); Peter, F.D.O.-A. 2: 103 (1932). Type: Kenya, Teita District, Ndi, *Hildebrandt* 2842 (BM, K, iso.!)

F. populifolia Vahl var. *major* Warb. in E.J. 36: 212 (1905); Peter, F.D.O.-A. 2: 103 (1932). Type: Tanzania, Mwanza District, Uzinza [Usindja], Ngama, *Stuhlmann* 3563(a) (B, ? holo., not found)

23. F. trichopoda *Bak.* in J.L.S. 20: 261 (1883); Troupin, Fl. Pl. Lign. Rwanda: 449, fig. 150.3 (1982); C.C. Berg et al. in Fl. Cameroun 28: 158, t. 52 (1985). Type: Madagascar, *Baron* 1663 (K, holo.!, B, iso.!)

Tree up to 10(–20) m. tall or a shrub, terrestrial, often with stilt or pillar roots. Leafy twigs 3–7 mm. in diameter, glabrous or white puberulous to hirtellous, periderm not flaking off. Leaves in spirals; lamina coriaceous, ± broadly ovate to elliptic, 6–20(–28) × 4–12(–21) cm., apex shortly acuminate to obtuse, base obtuse to cordate, margin entire; upper surface glabrous or puberulous to hirtellous on the main veins, lower surface white hirtellous to tomentellous, at least on the midrib, or sometimes glabrous; lateral veins 7–11 pairs, the basal pairs branched, reaching the margin below or sometimes at the middle of the lamina, tertiary venation partly scalariform; petiole 2–4(–7) cm. long, (1–)2–3 mm. thick, epiderm not flaking off; stipules 1.5–4.5(–8) cm. long, white puberulous to hirtellous, caducous. Figs up to 4 together in the leaf-axils; peduncle 0.5–1 cm. long; basal bracts ± 2 mm. long, persistent. Receptacle ± globose, 1–2 cm. in diameter when fresh, 0.5–1.5 cm. when dry, glabrous to ± densely puberulous, smooth or warted, red or yellow at maturity.

UGANDA. W. Nile District: Adumi, May 1936, *Eggeling* 3022!; Acholi District: Gulu, Mar. 1935, *Eggeling* 1642!; Mengo District: 120 km. Kampala–Masaka, July 1936, *Chandler & Hancock* 1774!

TANZANIA. Bukoba District: Bushasha, *Gillman* 336!; Rufiji District: Mafia I., Mwakuni, 7 Aug. 1937, *Greenway* 5012!; Songea District: R. Luhira N. of Songea, 23 June 1956, *Milne-Redhead & Taylor* 10888!

DISTR. **U** 1, 2, 4; **T** 1, 6–8; extending to Senegal and South Africa, also in Madagascar

HAB. Ground-water forest, swamp edges, riverine, wet valleys; 50–1200 m.

SYN. *F. congensis* Engl. in E.J. 8: 59 (1886); Hutch. in F.T.A. 6(2): 195 (1917); Peter, F.D.O.-A. 2: 104 (1932); Lebrun & Boutique in F.C.B. 1: 167 (1948); T.T.C.L.: 361 (1949); I.T.U., ed. 2: 245, fig. 56a (1952); F.W.T.A., ed. 2, 1: 609 (1958); F.F.N.R.: 33, t. 6L (1962); Hamilton, Ug. For. Trees: 98 (1981). Type: Zaire, Ponta da Lenha, *Naumann* 223 (B, holo.!)

F. budduensis Hutch. in K.B. 1915: 340 (1915) & in F.T.A. 6(2): 194 (1917). Type: Uganda, Masaka District, Buddu, *Dawe* 234 (K, holo.!)

24. F. lutea *Vahl*, Enum. Pl. 2: 185 (1805); Hutch. in F.T.A. 6(2): 215 (1917); Troupin, Fl. Pl. Lign. Rwanda: 445, fig. 149.3 (1982); C.C. Berg in K.B. 36: 597 (1981); C.C. Berg et al. in Fl. Cameroun 28: 206, t. 73 (1985). Type Ghana [Guinea], *Thonning* (no material found); neotype (see Berg, l.c.): Ghana, Aburi, *J. Hall* GC 47207 (U, neo.!, B, BM, K, isoneo.!)

Tree up to 20 m. tall, hemi-epiphytic or secondarily terrestrial, with a spreading crown. Leafy twigs 5–12(–20) mm. thick, puberulous, white to yellow tomentose to subhirsute or glabrous, periderm flaking off when dry. Leaves in spirals; lamina coriaceous, elliptic to oblong or ± obovate, 7–25(–45) × 3–12(–20) cm., apex acuminate, base obtuse to acute or ± cordate, margin entire; upper surface glabrous or puberulous on the midrib, lower surface sparsely to, less often, densely puberulous to hirtellous to ± tomentose, on the main veins to subhirsute, only subhirsute on the main veins, or entirely glabrous; lateral veins (4–)6–10(–12) pairs; tertiary venation partly scalariform; petiole 1.5–13(–17) cm. long, 2–3(–8) mm. thick, epiderm flaking off when dry; stipules 0.5–2.5 cm. long, on flush up to 8 cm., puberulous or also white to yellow subsericeous, caducous. Figs up to 4 together in the leaf-axils or ± just below the leaves, sessile, initially in a white pubescent to subhirsute calyptrate bud up to 5 mm. long; basal bracts 3–6 mm. long, persistent. Receptacle ± globose, ± 1–2.5(–? 3.5) cm. in diameter when fresh, 0.8–1.5(–2.5) cm. when dry, puberulous to white or yellow pubescent or subhirsute, smooth or warted, yellow to orange or brownish at maturity.

UGANDA. Bunyoro District: Budongo Forest, 12 Oct. 1962, *V. Reynolds* 9!; Masaka District: Nkose I., 21 Jan. 1956, *Dawkins* 865!

KENYA. Machakos District: Fourteen Falls, 6 Sept. 1952, *Verdcourt* 723! & Kibwezi (cult. Nairobi), *Bally* 12281!; N. Kavirondo District: Kakamega Forest, 7 Jan. 1968, *Perdue & Kibuwa* 9488!

TANZANIA. Masai District: Endulen, 24 Dec. 1962, *Newbould* 6459!; Moshi District: Rombo, 28 June 1946, *Greenway* 7833!; Rungwe District: Mwakaleli, 9 May 1975, *Hepper, Field & Mhoro* 5428!; Zanzibar I., Chwaka, 27 Sept. 1962, *Faulkner* 3104!

DISTR. **U** 2, 4; **K** 4, 5, 7; **T** 1–3, 6, 7; **Z**; **P**; extending to Ethiopia, Cape Verde Is., Angola, South Africa (Natal), Madagascar and the Seychelles

HAB. Forest, persisting in cleared areas and planted around settlements, riverine and by lakes; 0–1800 m.

SYN. *Urostigma luteum* (Vahl) Miq. in Lond. Journ. Bot. 6: 554 (1847)

U. vogelii Miq. in Lond. Journ. Bot. 7: 563, t. 12A (1848). Type: Liberia, Cape Palmas, *Vogel* 47 (U, lecto.!, K, isolecto.!)

Ficus vogelii (Miq.) Miq. in Ann. Mus. Lugd.-Bat. 3: 288 (1867); Hutch. in F.T.A. 6(2): 179 (1916); Lebrun & Boutique in F.C.B. 1: 146 (1948); T.T.C.L.: 360 (1949); U.O.P.Z.: 265 (1949); F.W.T.A., ed. 2, 1: 609 (1958); K.T.S.: 313 (1961); Aweke in Meded. Landb. Wageningen 79-3: 93, fig. 22 (1979)

F. quibeba Ficalho, Pl. Ut. Afr. Port.: 270 (1884); T.T.C.L.: 359 (1949); K.T.S.: 319 (1961). Type: Angola, Oueta Mts., *Welwitsch* 6399 (B, BM, K, P, iso.!)

F. subcalcarata Warb. & Schweinf. in E.J. 20: 155 (1894); Peter, F.D.O.-A. 2: 108 (1932); T.T.C.L.: 360 (1949); K.T.S.: 320 (1961). Type: Zaire, Monbutto, Munsa, *Schweinfurth* 3624 (B, holo.!, K, iso.!)

F. holstii Warb. in E.J. 20: 160 (1894); Peter, F.D.O.-A. 2: 108 (1932). Type: Tanzania, Lushoto District, Lutindi, *Holst* 3305 (B, holo.!)

F. lanigera Warb. in E.J. 20: 62 (1894). Type: Tanzania, Bukoba, *Stuhlmann* 1449 (B, holo.!)

F. verrucocarpa Warb. in E.J. 30: 294 (1901); Peter, F.D.O.-A. 2: 108 (1932); Lebrun & Boutique in F.C.B. 1: 157 (1948). Type: Tanzania, Rungwe District, Kiwira valley, *Goetze* 1492 (B, holo.!, K, iso.!)

F. nekbudu Warb. in Ann. Mus. Congo, Bot., sér 6, 1: 6 (1904); Lebrun & Boutique in F.C.B. 1: 145 (1948). Type: Zaire, Ubangi-Uele, Hakrakra, *Witterwulge* (B, holo.!, BR, iso.!)

F. vogelii (Miq.) Miq. var. *pubicarpa* Mildbr. & Burret in E.J. 46: 238 (1911). Lectotype: Togo, *Kersting* A539 (B, holo.!, K, iso.!)

F. stolzii Mildbr., *nom. nud.*, based on Tanzania, Rungwe District, Kondeland, *Stolz* 1531

NOTE. In West Africa *F. lutea* and *F. saussureana* are clear-cut species. In E. Zaire and Uganda, however, the differentiating characters (like dimension of the leaves and figs, the shape of the leaves and the number of lateral veins) are faint, so one wonders whether interbreeding may occur.

25. F. saussureana *DC.* in Mém. Soc. Phys. Hist. Nat., Genève 9: 2, t. (1841); C.C. Berg et al. in Fl. Cameroun 28: 209, t. 74 (1985). Type: plate in Mém Soc. Phys. Hist. Nat. Genève 9 (1841)

Tree up to 20 m. tall, hemi-epiphytic or secondarily terrestrial, with spreading crown. Leafy twigs 10–15 mm. in diameter, puberulous and usually also white to yellow pubescent or subhirsute, periderm flaking off when dry. Leaves in spirals; lamina coriaceous, oblong or subobovate or oblanceolate, (10–) 15–50 × 3–17(–25) cm., apex acuminate, base acute to obtuse or ± cordate, margin entire; upper surface glabrous, lower surface puberulous to hirtellous, on the main veins to subhirsute; lateral veins 12–20 pairs (in smaller blades, less than 15 cm. long, sometimes 10–11 pairs); tertiary venation partly scalariform; petiole 1–8(–16) cm. long, 4–7 mm. thick, periderm flaking off when dry; stipules 1–8 cm. long, on flush up to 12 cm., outside brown to yellowish pubescent, inside with long appressed yellow hairs, caducous or subpersistent. Figs up to 3 together in the leaf-axils or just below the leaves, subsessile, initially in a greyish subvillous calyptrate bud up to 1 cm. long; basal bracts 7–15 mm. long, persistent. Receptacle subglobose, sometimes obovoid, 2–4 cm. in diameter when fresh, 1.5–3 cm. when dry, yellow to orange-brown subhirsute to villous, yellow to orange or reddish at maturity.

UGANDA. W. Nile District: Zeu [Zeio], Mar. 1935, *Eggeling* 1882!; Bunyoro District: Siba Forest, Sept. 1962, *Reynolds* 6!; Masaka District: Bugala I., Feb. 1933, *A.S. Thomas* 827!
KENYA. N. Kavirondo District: Kakamega Forest, 24 Jan. 1982, *M.G. Gilbert* 6863!
TANZANIA. Mwanza District: Speke Gulf, May 1931, *B.D. Burtt* 2486! & Saanane I., Dec. 1951, *Tanner* 526!
DISTR. **U** 1, 2, 4; **K** 5; **T** 1; extending to S. Sudan and Guinée
HAB. Forest, often at edges, riverine or at lakesides; 900–1600 m.

SYN. *F. eriobotryoides* Kunth & Bouché in Ind. Sem. Hort. Berol. 1846: 14 (1847); Hutch. in F.T.A. 6(2): 160 (1916); Peter, F.D.O.-A. 2: 107 (1932); T.T.C.L.: 358 (1949); F.W.T.A., ed. 2, 1: 608 (1958). Type: Hort. Bot. Berol., 17 Sept. 1846 (B, holo.!), probably grown from material from Sierra Leone collected by Heesch
F. dawei Hutch. in K.B. 1915: 332, fig. (1915) & in F.T.A. 6(2): 161 (1916); I.T.U., ed. 2: 245 (1952). Type: Uganda, Masaka District, Buddu, *Dawe* 288 (K, holo.!)

NOTE. See note under *F. lutea*.

26. F. fischeri *Mildbr. & Burret* in E.J. 46: 227 (1911); Hutch. in F.T.A. 6(2): 126 (1916); Peter, F.D.O.-A. 2: 99 (1932); T.T.C.L.: 356 (1949); F.F.N.R.: 34 (1962). Type: Tanzania, Mwanza District, Kagehi, *Fischer* 545 (B, lecto.!)

Tree up to 15 m. tall, hemi-epiphytic, often soon terrestrial, with a flat-topped crown. Leafy twigs 4–10 mm. thick, glabrous or puberulous, periderm not flaking off. Leaves in spirals; lamina coriaceous, ovate to elliptic, (4–)6.5–17 × (3–)5–11 cm., apex acuminate, sometimes subacute or rounded, base cordate to truncate, sometimes rounded, margin entire; both surfaces glabrous; lateral veins 9–15 pairs; tertiary venation parallel to the lateral veins to reticulate; petiole 2.5–10 cm. long, (1–)1.5–3 mm. thick; stipules 0.3–0.8 cm. long, glabrous or puberulous, caducous. Figs solitary or in pairs in the leaf-axils; peduncle 0.8–1.8 cm. long; basal bracts 2–2.5 mm. long, caducous. Receptacle globose, 1.5–2 cm. in diameter when fresh, 1.5–2 cm. when dry, glabrous or minutely ± brown puberulous, yellowish green at maturity; wall of fruiting fig ± 2 mm. thick when dry, mostly ± wrinkled.

TANZANIA. Mwanza District: near Karumo, Mar. 1937, *B.D. Burtt* 6495!; Kondoa District: 12 km. E. of Kondoa, Jan. 1962, *Polhill & Paulo* 1234!; Mbeya District: Isunura, Sept. 1970, *Thulin & Mhoro* 1249!
DISTR. **T** 1, 4, 5, 7, 8; Zambia, Mozambique, Angola, Botswana, rarely in Malawi and Zimbabwe
HAB. Deciduous woodland and wooded grassland, sometimes planted for shade and bird-lime; 950–1500 m.

SYN. *F. kiloneura* Hornby in Bothalia 4: 1007 (1948). Type: Mozambique, Niassa, near Chiponde frontier post, *Hornby* 2471 (PRE, holo.!, K, iso.!)

27. F. amadiensis *De Wild.* in F.R. 12: 200 (1913); Lebrun & Boutique in F.C.B. 1: 150 (1948). Type: Zaire, Amadi, *Seret* 287 (BR, holo.!, K, iso.!)

Tree up to ± 15 m. tall, (? secondarily) terrestrial. Leafy twigs ± 5–10 mm. thick, white puberulous, periderm not flaking off. Leaves in spirals; lamina ± coriaceous, oblong to elliptic, ovate or lanceolate, sometimes ± obovate, (4–)7–17 × (2–)3–8.5 cm., apex obtuse to rounded, base obtuse to ± cordate, margin entire; both surfaces glabrous; lateral veins (8–)10–15 pairs; tertiary venation reticulate, but towards the midrib tending to parallel the lateral veins; petiole (1–)1.5–5.5(–7) cm. long, 1.5–2.5 mm. thick; stipules 0.5–1.5 cm. long, outside glabrous, puberulous or partly pubescent, inside in the lower part densely pubescent, subpersistent. Figs in pairs in the leaf-axils, sessile, initially in calyptrate buds up to ± 1 cm. long, splitting into 2 persistent scales, pubescent to puberulous outside, white pubescent to villous inside; basal bracts 1.5–2 mm. long, persistent. Receptacle ± globose, 1.2–2 cm. in diameter when dry; wall wrinkled when dry.

UGANDA. Acholi District: Lamogi, Mar. 1935, *Eggeling* 1657!; Toro District: Bubandi, Nov. 1935, *A.S. Thomas* 1472!; Mengo District: Port Kibanga, *Dummer* 2646!
KENYA. Trans-Nzoia District: Moi's [Hoey's] Bridge, Aug. 1963, *Heriz-Smith & Paulo* 907!; S. Kavirondo District: Bukuria [Bukeria], Sept. 1933, *Napier* 5273!; Kisumu-Londiani District: Songhor, Muhoroni valley, Nov. 1944, *S. Gillett* in *Bally* 4097!
TANZANIA. Biharamulo District: Nyabugombe, Nov. 1960, *Tanner* 5583A!; Rungwe District: Kyimbila, Aug. 1913, *Stolz* 2105!
DISTR. **U** 1–4; **K** 3, 5, 6; **T** 1, 7; Zaire, Rwanda
HAB. Wooded grassland, sometimes on hill tops or on ant-hills, or lakeside forest and thicket, rarely planted for bark-cloth; 950–2100 m.

SYN. *F. kitubalu* Hutch. in K.B. 1915: 334 (1915) & in F.T.A. 6(2): 163 (1916); I.T.U., ed. 2: 252 (1952); K.T.S.: 318 (1961); Troupin, Fl. Pl. Lign. Rwanda: 445 (1982). Type: Uganda, Masaka District, Buddu, *Dawe* 286 (K, holo.!)
F. calotropis Lebrun & Toussaint, Expl. Parc Nat. Kagera 1: 40 (1948); Lebrun & Boutique in F.C.B. 1: 158 (1948). Type: Rwanda, S. of Gabiro, Lugadzi, *Lebrun* 9521 (BR, holo.!, P, iso.!)
F. ndola Mildbr. in Willdenowia 1: 126 (1953). Type: Tanzania, Rungwe District: Kyimbila, *Stolz* 2105 (B, holo.!, BM, P, L, U, iso.!)

28. F. craterostoma *Mildbr. & Burret* in E.J. 46: 247 (1911); Hutch. in F.T.A. 6(2): 160 (1916); Peter, F.D.O.-A. 2: 108 (1932); T.T.C.L: 357 (1949); F.F.N.R.: 32, t. 6G (1962); C.C. Berg et al. in Fl. Cameroun 28: 182, t. 62 (1985). Type: Tanzania, Uluguru Mts., Ruvu R., *Stuhlmann* 8995 (B, lecto.!, K, isolecto.!)

Tree up to 10 m. tall or a shrub, hemi-epiphytic. Leafy twigs 2–5 mm. thick, glabrous or white puberulous to hirtellous, periderm sometimes flaking off when dry. Leaves in spirals, tending to distichous, often subopposite; lamina ± coriaceous, narrowly obtriangular to ± obovate, or oblong to elliptic, 3–8 × 2–4.5 cm., apex truncate to emarginate (or 2-lobed) or obtuse, base acute to obtuse, margin entire; both surfaces glabrous; lateral veins 5–10 pairs, midrib not reaching the apex of the lamina, tertiary venation reticulate or tending to parallel the lateral veins; petiole 0.5–2 cm. long, 1–2 mm. thick; stipules ± 0.5 cm. long, glabrous or yellowish to white puberulous, subpersistent or caducous. Figs in pairs in the leaf-axils, sessile, initially in a calyptrate bud up to 1 cm. long, splitting into 2 subpersistent or caducous parts, these pubescent inside; basal bracts 1–1.5 mm. long, persistent. Receptacle globose to ellipsoid, 0.8–1.2 cm. in diameter when fresh, ± 0.5 cm. in diameter when dry, glabrous or puberulous, reddish or sometimes yellowish at maturity; wall mostly slightly wrinkled when dry.

UGANDA. Kigezi District: Mpalo, July 1939, *Purseglove* 906! & Dec. 1934 , *C.M. Harris* 196!; Masaka District: Nkose I., Jan. 1956, *Dawkins* 849!
TANZANIA. Bukoba District: Biharamulo road, Nov. 1948, *Ford* 841! & Rubogo swamp, Sept.–Oct. 1935, *Gillman* 399!; Morogoro District: Uluguru Mts., Ruvu R., *Stuhlmann* 8995!
DISTR. **U** 2, 4; **K** 7; **T** 1, 3, ?4, 6; extending to Sierra Leone, Angola, N. Zambia and South Africa (Natal)
HAB. Forest, generally ground-water or riverine, occasionally planted as a fence; 300–2100 m.

SYN. *F. luteola* De Wild. in F.R. 12: 199 (1913); Hutch. in F.T.A. 6(2): 159 (1916); Lebrun & Boutique in F.C.B. 1: 142 (1948). Type: Zaire, Nala, *Seret* 801 (BR, syn.!, K, frag.!)
F. pilosula De Wild. in F.R. 12: 199 (1913); Lebrun & Boutique in F.C.B. 1: 142 (1948); I.T.U., ed. 2: 256, fig. 57c (1952). Type: Zaire, Eala, *Pynaert* 1130 (BR, holo.!, K, frag.!)
F. anomani Hutch. in K.B. 1915: 331 (1915); F.W.T.A., ed. 2, 1: 607 (1958). Type: Ghana, Sefwi, *Armitage* (K, lecto.!)

NOTE. This species is recorded from Zanzibar in U.O.P.Z.: 263 (1949) under the name *F. pilosula*, but no material has been seen from there.

29. F. natalensis *Hochst.* in Flora 28: 88 (1845); Hutch. in F.T.A. 6(2): 208 (1917) & in Fl. Cap. 5(2): 538 (1925); Peter, F.D.O.-A. 2: 104 (1932); T.T.C.L.: 361 (1949); U.O.P.Z.: 263 (1949); I.T.U., ed. 2: 253 (1952); F.W.T.A., ed. 2, 1: 610 (1958); K.T.S.: 319 (1961); F.F.N.R.: 34 (1962); Hamilton, Ug. For. Trees: 102 (1981); Troupin, Fl. Pl. Lign. Rwanda: 446, fig. 147.2 (1982); C.C. Berg et al. in Fl. Cameroun 28: 184, t. 63 (1985). Type: South Africa, Durban [Port Natal], *Krauss* (BM, K, U, isolecto.!)

Tree up to 30 m. tall or a shrub, hemi-epiphytic or (? secondarily) terrestrial. Leafy twigs 2–5 mm. thick, glabrous or sparsely and minutely puberulous, periderm not flaking off. Leaves in spirals to almost distichous and often subopposite; lamina ± coriaceous, oblong to elliptic or ± obovate, 2.5–10 × 1–4.5 cm., apex acuminate to obtuse, rounded or emarginate, base acute to obtuse, margin entire; both surfaces glabrous; lateral veins 6–13 pairs, midrib usually not reaching the apex of the lamina, tertiary venation reticulate to parallel to the lateral veins; petiole 0.5–2(–3) cm. long, 1–2(–2.5) mm. thick, glabrous;

stipules 0.2–1 cm. long, glabrous or sometimes minutely and sparsely puberulous, caducous. Figs in pairs in the leaf-axils or sometimes also just below the leaves, initially enclosed by a small or sometimes up to 1.5 cm. long almost glabrous calyptrate bud-cover; peduncle 0.2–1 cm. long; basal bracts 2–2.5 mm. long, caducous. Receptacle often shortly stipitate at least when dry, globose to ellipsoid or obovoid, ± 1.5–2 cm. in diameter when fresh, 0.8–1.5(–1.9) cm. when dry, glabrous, reddish, orange or yellowish (to brown) at maturity; wall rather thin, usually wrinkled when dry, apex plane or slightly protruding.

UGANDA. Bunyoro District: Biso [Bisu], May 1935, *Eggeling* 1985!; Masaka District: Buddu, Sango, 1905, *Dawe* 319!; Mengo District: Mawokota, Bussi I., Feb. 1932, *Eggeling* 199!
KENYA. Kiambu District: Ruiru, Dec. 1967, *Perdue & Kibuwa* 8197!; Nairobi, Oct. 1971, *Gillett* 19310!; Machakos District: Kibwezi, Dec. 1971, *Gillett* 19400!
TANZANIA. Moshi District: Rau Forest, Feb. 1953, *Drummond & Hemsley* 1309!; Tanga District: Muheza, Mar. 1956, *Faulkner* 1825!; Kondoa District: Bereku, Apr. 1974, *Richards* 29118!; Pemba I., Verani, Feb. 1929, *Greenway* 1460!
DISTR. **U** 2, 4; **K** 1, 4–7; **T** 1–3, 5–8; **Z**; **P**; extending to Senegal, S. Sudan, Angola, N. Zambia and to South Africa
HAB. Forest (both wet and dry types), groundwater and riverine forest, higher rainfall woodland, often planted for bark-cloth; 10–2200 m.

SYN. *Urostigma natalense* (Hochst.) Miq. in Lond. Journ. Bot. 6: 556 (1847)
Ficus volkensii Warb. in E.J. 20: 167 (1894); Hutch. in F.T.A. 6(2): 208 (1917). Type: Tanzania, Lushoto District, Derema [Nderema], *Volkens* 136 (B, holo.!, K, iso.!)

NOTE. This species can be easily confused with *F. thonningii*, *F. craterostoma* and *F. faulkneriana*. It can be distinguished from these three species by the pedunculate figs in conjunction with the caducous basal bracts. In West Africa (to Central Africa) the species is represented by subsp. *leprieurii* (Miq.) C.C. Berg (*F. leprieurii* Miq.) with small figs (± 5 mm. across when dry) and usually broadly obovate to obtriangular leaves.
Formerly the most important source of bark-cloth in Uganda, see I.T.U., ed. 2 (1952).

30. F. faulkneriana *C.C. Berg* in K.B. 43: 83 (1988). Type: Tanzania, Tanga District, Magunga Estate, Vigai, *Faulkner* 1168 (K, holo.!)

Tree up to 30 m. tall, (? secondarily) terrestrial. Leafy twigs 1.5–3 mm. thick, minutely puberulous, periderm not flaking off. Leaves in spirals, sometimes tending to subopposite; lamina ± coriaceous, obovate to elliptic, or sometimes oblong, 1.5–7 × 0.8–3.5(–4.5) cm., apex rounded to obtuse, sometimes very shortly and bluntly acuminate or emarginate, base rounded to cordulate or obtuse, margin entire; both surfaces glabrous; lateral veins (3–)5–8 pairs, midrib not reaching the apex of the lamina, tertiary venation reticulate or parallel to the lateral veins; petiole 0.2–1(–1.5) cm. long, 1–1.5 mm. thick, glabrous; stipules 0.2–0.6(–0.9) cm. long, sparsely puberulous or only ciliolate, often subpersistent. Figs in pairs in the leaf-axils or just below the leaves: peduncle (0.3–)0.5–1.5 cm. long, sparsely minutely puberulous; basal bracts 1.5–2 mm. long, persistent. Receptacle subglobose to obovoid, 0.4–0.8(–1.2) cm. long when dry, almost glabrous, reddish or yellowish at maturity.

KENYA. Kwale District: Lango ya Mwagandi [Longomwagandi], Feb. 1968, *Magogo & Glover* 51!
TANZANIA. Tanga District: Magunga Estate, Dec. 1953, *Faulkner* 1310!; Pangani District: Mwera, Aug. 1957, *Tanner* 3668! & Mkwaja, Mikocheni, Oct. 1957, *Tanner* 3719!
DISTR. **K** 7; **T** 3; not known elsewhere
HAB. Coastal bushland and wooded grassland; 0–450 m.

NOTE. This species is reminiscent of both *F. natalensis* and *F. thonningii*. It differs from *F. natalensis* in the persistent basal bracts and from the pedunculate forms of *F. thonningii* in the sparse and minute hairs on the twigs, stipules and peduncle and/or the glabrous lamina and petiole. The species is probably most closely related to *F. lingua*.

31. F. lingua *De Wild. & Th. Dur.* in Ann. Mus. Congo, Bot., sér. 3, 2: 216 (1901); Warb. in Ann. Mus. Congo, Bot., sér. 6, 1: 24 (1904); Hutch. in F.T.A. 6(2): 156 (1916); Lebrun & Boutique in F.C.B. 1: 138 (1948); F.W.T.A., ed. 2, 1: 608 (1958); C.C. Berg et al. in Fl. Cameroun 28: 180, t. 61 (1985). Type: Zaire, Lokandu, near La Lowa, *Dewèvre* 1136 (BR, holo.!, B, iso.!)

Tree up to 30 m. tall, hemi-epiphytic or (secondarily) terrestrial. Leafy twigs 1–4 mm. thick, whitish or brownish puberulous, periderm of older parts peeling off when dry. Leaves in spirals, tending to distichous; lamina ± coriaceous, oblanceolate to subobovate or narrowly obtriangular, 0.5–5 × 0.3–2(–3) cm., apex obtuse to truncate, emarginate or subacute, base cuneate to obtuse, margin entire; both surfaces glabrous; lateral veins 5–8 pairs; tertiary venation reticulate or tending to parallel the lateral veins; petiole 0.2–0.8 cm. long, 0.5–1 mm. thick; stipules 0.2–0.5 cm. long, puberulous or only ciliolate, caducous, sometimes subpersistent. Figs in pairs in the leaf- axils or just below the leaves, on peduncles 2–5 mm. long or sometimes subsessile; basal bracts 1–2 mm. long, persistent or caducous. Receptacle globose or sometimes ellipsoid, ± 5 mm. in diameter when fresh, 3–4 mm. when dry, minutely puberulous, reddish or yellowish at maturity; wall thin, smooth or slightly wrinkled when dry.

subsp. **lingua**

Leafy twigs with brownish hairs; stipules usually persistent. Peduncle usually glabrous; basal bracts caducous.

UGANDA. Bunyoro District: Budongo Forest, July 1935, *Eggeling* 2085! & July 1936 *Eggeling* 3080!; Masaka District: Minziro Forest, July 1938, *Eggeling* 3740!
DISTR. U 2, 4; extending to Cameroun, also in Ivory Coast and (*fide* F.W.T.A.) Liberia
HAB. Forest: 1050–1200 m.

SYN. [*F. depauperata* sensu I.T.U., ed. 2: 247 (1952), *non* Sim sensu stricto]

subsp. **depauperata** (*Sim*) *C.C. Berg* in K.B. 43: 85 (1988). Type: Mozambique, *Sim* 5031 (not found)

Leafy twigs with whitish hairs; stipules caducous. Peduncle usually minutely puberulous; basal bracts persistent.

KENYA. Kwale District: Diani Forest, July 1972, *Gillett & Kibuwa* 19904!; Kilifi District: 7 km. E. of Jilore Forest Station, Nov. 1969, *Perdue & Kibuwa* 10103!; Lamu District: Utwani Forest Reserve, Mambasasa, Oct. 1957, *Greenway & Rawlins* 9346!
TANZANIA. Uzaramo District: Dar es Salaam, Dec. 1967, *Harris* 1262!; Rufiji District: Mafia I., Sept. 1937, *Greenway* 5262!; Pemba I., Kwata I., Dec. 1930, *Greenway* 2738!
DISTR. **K** 7; **T** 6; **P**; Mozambique
HAB. Forest, coastal bushland, coral outcrops; 0–50 m.

SYN. *F. depauperata* Sim, For. Fl. Port. E. Afr.: 98, t. 40 (1909); Hutch. in F.T.A. 6(2): 204 (1917); Peter, F.D.O.-A. 2: 106 (1932); T.T.C.L.: 361 (1949); U.O.P.Z.: 262 (1949); K.T.S.: 316 (1961)

32. F. thonningii *Bl.* in Rumphia 2: 17 (1836); Hutch. in F.T.A. 6(2): 187 (1916); Peter, F.D.O.-A. 2: 109 (1932); Lebrun & Boutique in F.C.B. 1: 148, t. 16 (1948); T.T.C.L.: 360 (1949); I.T.U., ed. 2: 260 (1952); F.P.S. 2: 270 (1952); F.W.T.A., ed. 2, 1: 610 (1958); K.T.S.: 321 (1961); Hamilton, Ug. For. Trees: 104(1981); C.C. Berg in Fl. Cameroun 28: 175, t. 59 (1985). Type: Ghana, *Thonning* 325 (not traced, probably lost)

Tree up to 15(–30) m. tall or a shrub, terrestrial or hemi-epiphytic. Leafy twigs 1.5–7 mm. thick, minutely puberulous to hirtellous or white to brown pubescent, at least on the scars of the stipules or sometimes entirely glabrous, periderm usually not flaking off. Leaves in spirals, occasionally subopposite; lamina ± coriaceous, elliptic to lanceolate, oblanceolate, obovate or subovate, (1.5–)3–12(–18) × (1–)1.5–6(–7) cm., apex acuminate to obtuse or rounded, base subacute to rounded or subcordate, margin entire; upper surface glabrous or sparsely (on the midrib to rather densely) puberulous to pubescent, lower surface glabrous or sparsely to densely white (to brownish) puberulous to pubescent on the whole surface, the main veins or only the midrib; lateral veins (5–)7–12(–16) pairs, midrib often reaching the apex of the lamina (even in leaves with a rounded apex); tertiary venation reticulate or parallel to the lateral veins; petiole (0.5–)1–4(–6) cm. long, 1–2 mm. thick, often (not depending on the size of the lamina or the position of the leaf on the twig) variable in length on the same twig, glabrous or puberulous, hirtellous or pubescent; stipules 0.3–1(–1.5) cm. long, white to brown pubescent, puberulous or only ciliolate, caducous or subpersistent. Figs in pairs in the leaf-axils or sometimes also below the leaves, sessile or on peduncles up to 1 cm. long; basal bracts 2–4 mm. long, persistent. Receptacle globose to ellipsoid, 0.5–1.5(–2) cm. in diameter when fresh, 0.4–1.2(–1.7) cm. when dry, glabrous or sparsely to densely white to brown puberulous or pubescent, reddish, yellowish or brownish at maturity; wall thin, mostly smooth or slightly wrinkled when dry; apex plane to strongly protruding when dry. Fig. 21, p.74.

UGANDA. W. Nile District: Kango, May 1936, *Eggeling* 3020!; Kigezi District: Kachwekano Farm, Dec. 1951, *Purseglove* 3745!; Masaka District: Lake Nabugabo, 9 Oct. 1953, *Drummond & Hemsley* 4721!
KENYA. Nandi District: Kaimosi, 15 June 1953, *G.R. Williams* 578!; Machakos District: Kilungu Forest, 9 Jan. 1972, *Mwangangi* 1948!; Kisumu-Londiani District: Tinderet Forest Reserve, 1 July 1949, *Maas Geesteranus* 5312!

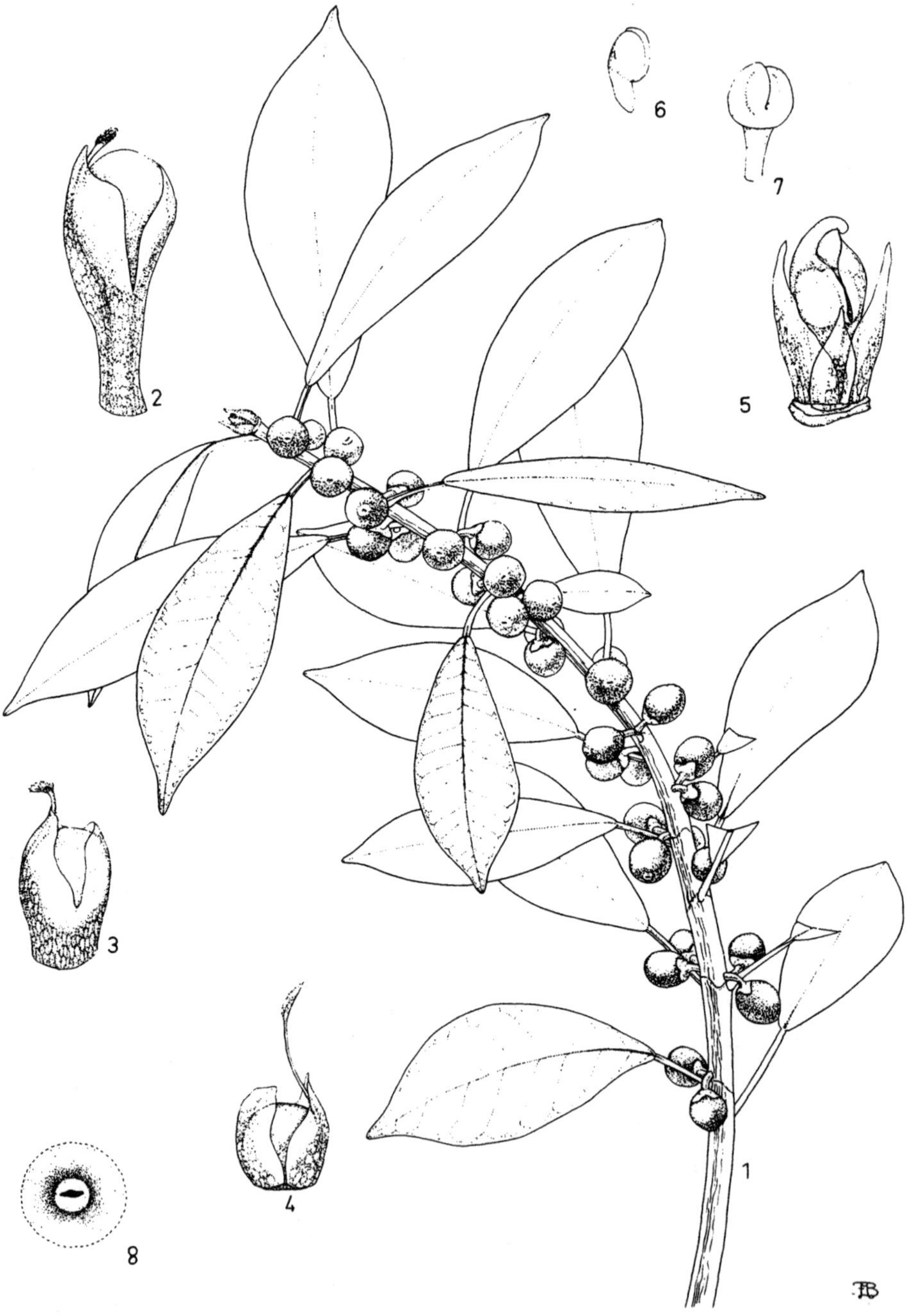

FIG. 21. *FICUS THONNINGII* — **1**, leafy twig with figs; **2**, **3**, gall flowers; **4**, seed flower; **5**, staminate flower; **6**, **7**, stamens; **8**, ostiole. All from *Aweke & Gilbert* 826. Drawn by F.M. Bata-Gillot.

TANZANIA. Arusha District: Ngurdoto Crater Rim, 10 Apr. 1968, *Greenway & Kanuri* 13442!; Dodoma District: Great North Road near Meia Meia, 25 Jan. 1962, *Polhill & Paulo* 1259!; Songea District: 17 km. W. of Songea, 24 Feb. 1956, *Milne-Redhead & Taylor* 8881!

DISTR. **U** 1–4; **K** 1–7; **T** 1–8; extending to Cape Verde Is., Angola, Ethiopia and South Africa

HAB. Forest, woodland, bushland and woodedd grassland, sometimes along rivers and lakes or among rocks, planted for ornament and bark-cloth; 350–2500 m.

SYN. *F. microcarpa* Vahl, Enum. Pl. 2: 188 (1805), *non* L.f. (1781), *nom. illegit.* Type as for species

Urostigma burkei Miq. in Hook., Lond. Journ. Bot. 6: 555 (1847). Type: South Africa, Transvaal, Magaliesberg, *Burke* (K, holo.!)

U. hochstetteri Miq. in Hook., Lond. Journ. Bot. 6: 555 (1847). Type: Ethiopia, Schagenni Region, *Schimper* 373 (U, holo.!, B, BR, K, L, P, iso.!)

U. schimperi Miq. in Hook., Lond. Journ. Bot. 6: 555, t. 22A (1847). Type: Ethiopia, Haramat Mts., near Geraz, *Schimper* 1096 (L, holo.!, B, BM, P, iso.!)

U. thonningii (Bl.) Miq. in Hook., Lond. Journ. Bot. 6: 558 (1847)

U. dekdekena Miq. in Hook., Lond. Journ. Bot. 6: 558 (1847). Type: Ethiopia, Mt. Sholoda, *Schimper* 220 (L, lecto.!, BM, K, P, isolecto.!)

Ficus hochstetteri (Miq.) A. Rich., Tent. Fl. Abyss. 2: 267 (1851); Peter, F.D.O.-A. 2: 112 (1932); T.T.C.L.: 359 (1949); Aweke in Meded. Landb. Wageningen 79-3: 33, fig. 8 (1979)

F. schimperi (Miq.) A. Rich., Tent. Fl. Abyss. 2: 267 (1851); Hutch. in F.T.A. 6(2): 188 (1966)

F. dekdekena (Miq.) A. Rich., Tent. Fl. Abyss. 2: 268 (1851); Hutch. in F.T.A. 6(2): 211 (1917); Peter, F.D.O.-A. 2: 106 (1932); T.T.C.L.: 361 (1949); I.T.U., ed. 2: 247 (1952); F.P.S. 2: 272 (1952); F.W.T.A., ed. 2, 1: (1958); K.T.S.: 316 (1961); F.F.N.R.: 34 (1962); Troupin, Fl. Pl. Lign. Rwanda: 442, fig. 147.5 (1982)

F. burkei (Miq.) Miq. in Ann. Mus. Bot. Lugd.-Bat. 3: 289 (1867); Hutch. in F.T.A. 6(2): 202 (1917); F.F.N.R.: 34 (1962); Troupin, Fl. Pl. Lign. Rwanda: 442, fig. 150.5 (1982)

F. persicifolia Warb. in E.J. 20: 162 (1894); P.O.A. C: 162, t. 8G–J (1895); Peter, F.D.O.-A. 2: 106 (1932); T.T.C.L.: 361 (1949); I.T.U., ed. 2: 256, fig. 57b (1952); F.F.N.R.: 33 (1962); Hamilton, Ug. For. Trees: 104 (1981). Type: Angola, Golungo Alto, *Welwitsch* 6337 (BM, C, P, iso.!)

F. chlamydodora Warb. in E.J. 20: 163 (1894); P.O.A. C: 161, t. 8A–F (1895); Hutch. in F.T.A. 6(2): 188 (1916). Type: Tanzania, Tabora, *Stuhlmann* 505 (B, holo.!)

F. petersii Warb. in E.J. 20: 164 (1894); Peter, F.D.O.-A. 2: 109 (1932); T.T.C.L.: 359 (1949); F.F.N.R.: 34 (1962). Type: Mozambique, Moravi, *Peters* (B, holo.!)

F. rokko Warb. & Schweinf. in E.J. 20: 164 (1894); Hutch. in F.T.A. 6(2): 188 (1916). Lectotype: Sudan, Niamniam, Ubangi, *Schweinfurth* 3038 (B, holo.!)

F. mabifolia Warb. in E.J. 20: 165 (1894); Hutch. in F.T.A. 6(2): 188 (1916). Type: Tanzania, Mwanza District, Busisi Creek, *Stuhlmann* (B, holo.!)

F. medullaris Warb. in E.J. 20: 169 (1894); Hutch. in F.T.A. 6(2): 188 (1916). Type: Zaire, Kivu, Kirima, *Stuhlmann* 2277a (B, holo.!)

F. goetzei Warb. in E.J. 28: 378 (1900); Hutch. in F.T.A. 6(2): 188 (1916). Type: Tanzania, Iringa District, Uzungwa [Utschungwe] Mts., near Muhanga, *Goetze* 662 (B, holo.!)

F. ruficeps Warb. in E.J. 30: 294 (1901); Hutch. in F.T.A. 6(2): 182 (1916). Type: Tanzania, Mbeya District, Usafwa, near Utengule, *Goetze* 1057 (B, holo.!, L, iso.!)

F. eriocarpa Warb. in E.J. 30: 294 (1901); Peter, F.D.O.-A. 2: 112 (1932); T.T.C.L.: 358 (1949); K.T.S.: 316 (1961). Type: Tanzania, Mbeya/Chunya Districts, Usafwa, near Swira, *Goetze* 1348 (B, holo.!, BM, K, iso.!)

F. ruspolii Warb. in E.J. 36: 211 (1905); Aweke in Meded. Landb. Wageningen 79-3: 60, fig. 15 (1979). Type: Ethiopia, Ciaffa, Boran Uata, *Ruspoli & Riva* 526 (FT, holo., K, iso.!)

F. rhodesica Mildbr. & Burret in E.J. 46: 254 (1911); Hutch. in F.T.A. 6(2): 201 (1917); I.T.U., ed. 2: 258 (1952); F.F.N.R.: 34 (1962). Type: Zimbabwe, Harare [Salisbury], *Engler* 3060 (B, holo.!, K, iso.!)

F. cyphocarpa Mildbr. in E.J. 46: 261 (1911); Peter, F.D.O.-A. 2: 106 (1932); Lebrun & Boutique in F.C.B. 1: 164 (1948). Type: Zaire, Beni–Mwera, *Mildbraed* 2392 (B, holo.!)

F. mammigera R.E. Fries in N.B.G.B. 8: 669 (1924); K.T.S.: 318 (1961). Type: Mt. Kenya, W. Kenya Forest Station, *Fries* 913 (UPS, holo.)

F. dekdekena (Miq.) A. Rich. var. *angustifolia* Peter, F.D.O.-A.: 106 (1932); T.T.C.L.: 361 (1949). Type: Tanzania, Tanga District, near Tengeni, *Peter* 16717 (not found)

F. thonningii Bl. var. *heterophylla* Peter, F.D.O.-A. 2: 111, Descr.: 9, t. 2/2. 2 (1932); T.T.C.L.: 360 (1949). Types: Tanzania, Arusha District, Lengijawe [Longidjawa], *Peter* 42668 (not found) & Masai District, Magadi, Olbussare stream, *Peter* 43388 (not found)

F. neurocarpa Lebrun & Toussaint in Expl. Parc Nat. Kagera 1: 42 (1948) & in F.C.B. 1: 171 (1948); Troupin, Fl. Pl. Lign. Rwanda: 446 (1982). Type: Rwanda, Mt. Lutare, *Lebrun* 9596 (BR, holo.!)

NOTE. The material treated under *F. thonningii* shows a confusing variation. The extremes of the variation are strikingly different, so that they look like distinct taxa. *F. thonningii* looks like a species complex or a complex in speciation. To handle the variation informal entities for the extremes of the variation are recognised. These entities ("forms") are more or less clearly associated with geography and habitat.

a. "**persicifolia**" form

Leafy twigs usually densely brown puberulous to pubescent. Leaves drooping (?); lamina mostly oblanceolate to oblong, often drying brown; midrib reaching the apex of the lamina; petiole relatively long and slender; stipules often subpersistent. Figs sessile to shortly pedunculate, 0.5–0.8 cm. in diameter when dry, almost glabrous; apex not or slightly protruding when dry.

DISTR. **U** 2–4; **T** 6, 7
HAB. Generally in forest, along lakes or by rivers, rarely in wooded grassland; 900–1750 m.

b. "**petersii**" form

In general features similar to the "*persicifolia*" form, but the lamina often elliptic to obovate, often drying greenish; stipules mostly caducous. Figs sessile or subsessile, 0.8–1 cm. in diameter when dry, with dense brown indumentum; apex not or only slightly protruding when dry.

DISTR. **K** 4; **T** 2, 5, 8
HAB. Deciduous woodland and bushland, rocky places; 350–1350 m.

c. "**dekdekena**" form

Largely similar to the "*petersii*" form, but the figs sessile or pedunculate and almost glabrous.

DISTR. **U** 1, 3; **T** 1
HAB. Wooded grassland, riverine and lakesides; 1050–2100 m.

d. "**burkei**" form

Leafy twigs mostly densely whitish (to pale brown) puberulous to pubescent, the same indumentum on the lower leaf-surface, the petiole and/or figs; lamina elliptic to obovate or oblong to subobovate; midrib reaching the apex of the lamina or not; petiole relatively short and stout. Figs subsessile to pedunculate, ± 0.4–1.1 cm. in diameter when dry; apex not or only slightly protruding when dry.

DISTR. **U** 1, 4; **K** 4, 6; **T** 1, 2, 4, 5, 7, 8
HAB. Wooded grassland, thickets; 900–1800 m.

e. "**neurocarpa**" form

Leafy twigs, as well as the petiole, ± densely puberulous to hirtellous; lamina mostly glabrous, elliptic to oblong, up to 6.5 × 3.5 cm.; midrib usually not reaching the apex of the lamina; petiole short, up to 1(–1.5) cm. long, stout; stipules caducous. Figs on peduncles 0.2–0.5 cm. long, 0.5–0.8 cm. in diameter when dry, sparsely to densely puberulous to pubescent; apex not or slightly protruding when dry.

DISTR. **U** 2; **T** 7
HAB. Rocky places, thickets; 1500–1800 m.

f. "**mammigera**" form

Leafy twigs, as well as other parts, sparsely puberulous to almost glabrous; stipules caducous. Figs sessile or subsessile, 0.5–1.2(–1.7) cm. in diameter when dry; apex ± protruding when dry.

DISTR. **U** 1–3; **K** 2–6; **T** 2, 3, 5, 7
HAB. Forest, riverine, sometimes in wooded grassland and bushland; 1000–2300 m.

33. F. ottoniifolia (*Miq.*) *Miq.* in Ann. Mus. Lugd.-Bat. 3: 288 (1867); Hutch. in F.T.A. 6(2): 134 (1916); Lebrun & Boutique in F.C.B. 1: 134 (1948); F.W.T.A., ed. 2, 1: 611 (1958); Hamilton, Ug. For. Trees: 99 (1981); Troupin, Fl. Pl. Lign. Rwanda: 466 fig. 147.3 (1982); C.C. Berg in Fl. Cameroun 28: 217, t. 77 (1985). Type: Fernando Po, *Vogel* 176 (K, holo.!)

Tree up to 15 m. tall or sometimes a shrub or liana, terrestrial or hemi-epiphytic. Leafy twigs 2–5(–10) mm. thick, minutely puberulous or glabrous, periderm not flaking off. Leaves in spirals; lamina coriaceous or sometimes chartaceous, elliptic to oblong, ovate or obovate, 6.5–15(–22) × 3–7(–9) cm., apex acuminate to subcaudate, base acute to rounded or subcordate, margin entire; both surfaces glabrous; lateral veins 5–12(–14) pairs, tertiary venation reticulate; petiole 1.5–9 cm. long (0.5–)1.5–2 mm. thick; stipules 0.2–0.8 cm. long, up to 4 cm. on flush, sparsely puberulous or glabrous, caducous. Figs up to 4(–10) together on spurs up to 1.5(–4) cm. long on the older wood; peduncle 0.8–2.5 cm. long, 1–1.5 mm. thick; basal bracts 2–3 mm. long, free parts caducous or subpersistent. Receptacle ellipsoid to globose, 1.5–2.5 cm. in diameter when fresh, 0.8–2 cm. when dry, puberulous or almost glabrous, greenish to pale orange or brownish at maturity, with pale green to whitish spots; wall ± 1 mm. thick when dry, not or hardly wrinkled.

KEY TO INFRASPECIFIC VARIANTS

Basal bracts caducous; lateral veins 3–10 pairs:
Lateral veins 6–10 pairs; receptacle 2–2.5 cm. in diameter when fresh, 1–1.5 cm. when dry . . . . a. subsp. **ottoniifolia**
Lateral veins 3–8 pairs; receptacle 1.5 cm. in diameter when fresh, 0.8–1.2 cm. when dry . . . . b. subsp. **lucanda**
Basal bracts ± persistent; lateral nerves 8–14 pairs . . . . c. subsp. **ulugurensis**

a. subsp. **ottoniifolia**

Lamina ± coriaceous, 8–22 × 3–9 cm., apex acuminate, base rounded to obtuse; lateral veins 6–10 pairs; petiole 2–13.5 cm. long, 1–2 mm. thick when dry. Peduncle 1–2.5 cm. long, 1–2 mm. thick when dry; basal bracts caducous. Receptacle ellipsoid to subglobose, 2–2.5 cm. in diameter when fresh, 1–1.5 cm. when dry, pale orange-green to pale yellow at maturity, with paler green to white spots.

UGANDA. Ankole District: Lutoto, 14 Sept. 1941, *A.S. Thomas* 3981!
DISTR. **U** 2; extending to Sierra Leone
HAB. Collected in Uganda only once in lakeside forest at 1350 m., elsewhere in rain-forest and drier types of evergreen forest.

SYN. *Urostigma ottoniifolium* Miq. in Hook., Lond. Journ. Bot. 7: 536, t. 13B (1848)

b. subsp. **lucanda** (*Ficalho*) *C.C. Berg* in K.B. 43: 90 (1988). Type: Angola, Golungo Alto, *Welwitsch* 6392 (LISU, holo., B, K, P, iso.!)

Lamina ± coriaceous, 4.5–15 × 2–7 cm., apex acuminate to subcaudate, base acute to obtuse, sometimes almost rounded; lateral veins (3–)4–8 pairs; petiole 1.5–7 cm. long. Peduncle 0.5–2.5 cm. long, 1–1.5 mm. thick; basal bracts caducous. Receptacle ellipsoid to subglobose, 1.5–2 cm. in diameter when fresh, 0.8–1.2 cm. when dry.

UGANDA. Ankole District: Lutoto, Oct. 1940, *Eggeling* 4103!; Kigezi District: Kayonza, Apr. 1948, *Purseglove* 2655!; Masaka District: S. Buddu, 1905, *Dawe* 296!
TANZANIA. Kigoma District: Gombe Stream Reserve, Nyasanga valley, 9 June 1970, *Clutton-Brock* 670 (in part)!
DISTR. **U** 2, 4; **T** 1, 4; Zaire, Gabon and N. Angola
HAB. Rain-forest and drier types of evergreen forest, riverine; 900–1500 m.

SYN. *F. lucanda* Ficalho in Pl. Ut. Afr. Port.: 269 (1884); Hutch. in F.T.A. 6(2): 133 (1916) as "*lukanda*"; I.T.U., ed. 2: 252 (1952)
F. sterculioides Warb. in E.J. 20: 175 (1894). Type: Tanzania, Bukoba, *Stuhlmann* 1019 (B, holo.!)

c. subsp. **ulugurensis** (*Mildbr. & Burret*) *C.C. Berg* in K.B. 43: 92 (1988). Type: Tanzania, Uluguru Mts., Mbora, *Stuhlmann* 2022 [not 9022] (B, lecto.!)

Lamina ± coriaceous, 7.5–15 × 3–7 cm., apex shortly acuminate, base subacute to subcordate; lateral veins 8–12(–14) pairs; petiole 1.5–4(–4.5) cm. long, 1–1.5 mm. thick when dry. Peduncle 0.8–1.8(–2) cm. long, 1–1.5 mm. thick when dry; basal bracts ± persistent. Receptacle ellipsoid to subglobose, 2–2.5 cm. in diameter when fresh, (1–)1.2–1.8 cm. when dry.

UGANDA. Kigezi District: Bwimbi [Impenetrable] Forest, Nyebeya, Oct. 1940, *Eggeling* 4163!
KENYA. Kwale District: Twiga, 18 Jan. 1964, *Verdcourt* 3949!; Kilifi District: Chasimba, 24 Nov. 1974, *B.R. Adams* 116!
TANZANIA. Lushoto District: Mavumbi Peak, 7 July 1953, *Drummond & Hemsley* 3196! & Amani, 7 Aug. 1930, *Greenway* 2375!
DISTR. **U** 2; **K** 7; **T** 3, 6, 7; not known elsewhere
HAB. Rain-forest, riverine, coastal bushland; 0–1500 m.

SYN. *F. ulugurensis* Mildbr. & Burret in E.J. 46: 226, fig. 4 (1911); Peter, F.D.O.-A. 2: 100 (1932); T.T.C.L.: 357 (1949)
F. scheffleri Mildbr. & Burret in E.J. 46: 225 (1911); Peter, F.D.O.-A. 2: 100 (1932); T.T.C.L.: 356 (1949). Type: Tanzania, Usambara Mts., Derema, *Scheffler* 215 (B, holo.!, BM, K, iso.!)

NOTE. Subsp. *macrosyce* C.C. Berg, lianescent with large figs 2–2.5 cm. in diameter, occurs in Zambia, southern Zaire and NE. Angola.

34. F. tremula *Warb.* in E.J. 20: 171 (1894); P.O.A. C: 162, fig. 10F–K (1895); Hutch. in F.T.A. 6(2): 137 (1916); Peter, F.D.O.-A. 2: 101 (1932); T.T.C.L.: 357 (1949); U.O.P.Z.: 265 (1949). Type: Tanzania, Bagamoyo, *Stuhlmann* 274 (B, holo.!)

Tree up to 10 m. tall or a shrub, hemi-epiphytic (and strangling) or (? secondarily) terrestrial, sometimes a liana. Leafy twigs 1–3(–5) mm. thick, sparsely minutely puberulous, periderm not flaking off. Leaves in spirals; lamina subcoriaceous to chartaceous, oblong to subobovate, elliptic or ± ovate, 2.5–8(–11) × 0.7–4(–5) cm., apex subacute to rather faintly acuminate, base rounded to cordulate, margin entire; both surfaces glabrous or the midrib puberulous beneath; lateral veins 5–9 pairs, tertiary venation reticulate; petiole 0.7–3(–4.5) cm. long, ± 0.5(–1) mm. thick; stipules 0.2–1 cm. long, up to 3 cm. on flush, glabrous, caducous. Figs 1–6 together on ± curved spurs up to 3 cm. long on the older wood; peduncle 0.5–2 cm. long, 0.5–7 mm. thick; basal bracts ± 3 mm. long, free parts caducous, occasionally subpersistent. Receptacle subglobose to ellipsoid, 2–2.5(?–3) cm. in diameter when fresh, 1–1.5(–2) cm. when dry, sparsely to sometimes densely minutely puberulous to almost glabrous, ? green at maturity, when dry not or hardly wrinkled and slightly stipitate; wall thin.

subsp. **tremula**

Tree or sometimes a climber. Twigs usually drying yellowish or greyish. Lamina mostly drying dark brown above and greenish beneath, base rounded to emarginate.

KENYA. Kwale District: W. of Kwale, Tanga–Mombasa road, 25 Jan. 1961, *Greenway* 9800! & Twiga, 14 Oct. 1962, *Verdcourt* 3288! & Diani Forest, 11–13 July 1972, *Gillett & Kibuwa* 19893!
TANZANIA. Tanga District: Mtotohovu, 12 Nov. 1947, *Brenan & Greenway* 8315! & 10 Sept. 1951, *Greenway* 8707!; Rufiji District: Mafia I., 3 Apr. 1933, *Wallace* 752!; Pemba, Kwata I., 16 Dec. 1930, *Greenway* 2739!
DISTR. **K** 7; **T** 3, 6; **Z**; **P**; Mozambique, E. Zimbabwe, South Africa (Natal)
HAB. Lowland dry evergreen forest, woodland and coastal bushland; 0–600 m.

SYN. *F. pulvinata* Warb. in E.J. 20: 169 (1894). Type: Tanzania, Zanzibar, *Stuhlmann* I–110 (B, holo.!)

subsp. **acuta** (*De Wild.*) *C.C. Berg* in K.B. 43: 96 (1988). Type: Zaire, Kivu, Mukule, *Bequaert* 6313 (BR, holo.!)

Tree or often a liana. Twigs drying brown to blackish. Lamina drying brownish, without a strong colour contrast between the upper and lower surfaces, base obtuse to rounded, rarely subcordate.

UGANDA. Kigezi District: Impenetrable Forest, Nyebeya, Oct. 1940, *Eggeling* 4163!
KENYA. Kericho District: 10 km. E. of Kericho, Sambret Estate, 6 Dec. 1967, *Perdue & Kibuwa* 9284! & Sambret–Timbilil, Oct. 1961, *Kerfoot* 2951! & SW. Mau Forest Reserve, 15 Aug. 1949, *Maas Geesteranus* 15801!
DISTR. **U** 2; **K** 5; E. Zaire, Rwanda and Burundi
HAB. Upland rain-forest; 1650–2300 m.

SYN. *F. acuta* De Wild. in Ann. Soc. Sci. Brux. 40: 278 (1921)

NOTE. The morphological differences between these two subspecies and a third, subsp. *kimuenzensis* (Warb.) C.C. Berg, in lowland western Africa, are small, partly only differences in the variation pattern. Geographical separation and differences in ecology appear to justify the distinction of the subspecies.

35. F. artocarpoides *Warb.* in Ann. Mus. Congo, Bot., sér. 6, 1: 23, t. 3 (1904); Hutch. in F.T.A. 6(2): 132 (1916); Lebrun & Boutique in F.C.B. 1: 136 (1948); Hamilton, Ug. For. Trees: 102 (1981); C.C. Berg et al. in Fl. Cameroun 28: 216, t. 76 (1985). Type: Zaire, *Gillet* 2014 (B, holo.!, BR, iso.)

Tree up to 10(–?15) m. tall, hemi-epiphytic (and strangling). Leafy twigs 2–4 mm. thick, minutely puberulous, periderm not flaking off. Leaves in spirals; lamina coriaceous, oblong to subobovate or oblanceolate, 6–21 × 2–6 cm., apex subacute to obtuse or sometimes rounded, base obtuse to acute, margin entire; upper surface glabrous, lower surface minutely puberulous on the main veins; lateral veins 10–16 pairs, tertiary venation reticulate or parallel to the lateral veins; petiole 1–3.5 cm. long, (1–)2 mm. thick; stipules 0.3–4 cm. long, puberulous or glabrous, caducous. Figs up to 5 together on ± curved spurs up to 4 cm. long and ± 1 cm. thick on the main and lesser branches (or just below the leaves); peduncle 1.5–3.5 cm. long, ± 2 mm. thick; basal bracts with free parts caducous. Receptacle subglobose to globose, 3–4 cm. in diameter when fresh, 2–3.5 cm. when dry, minutely puberulous, greenish to purplish at maturity; wall 1–1.5 mm. thick when dry and wrinkled.

UGANDA. Mengo District: Kiwala, May 1917, *Dummer* 3196! & Entebbe, Nambigirwa, Jan. 1932, *Eggeling* 142! & Kiagwe, Namanve [Niamanve], Oct. 1932, *Eggeling* 1528!
TANZANIA. Mpanda District: Mahali Mts., 27 Oct. 1987, *Nishida* 8702!
DISTR. **U** 4; **T** 4; extending to Ivory Coast and N. Angola

HAB. Rain-forest and riverine forest, elsewhere drier types of evergreen forest, also in secondary associations; 800–1200 m.

SYN. *Urostigma elegans* Miq. in Hook., Lond. Journ. Bot. 7: 563, fig. 13A (1848). Type: Ghana, Cape Coast, *Vogel* 87 (U, holo.!, ? K, iso.!)
Ficus elegans (Miq.) Miq. in Ann. Mus. Lugd.-Bat. 3: 268 (1867); Hutch. in F.T.A. 6(2): 128 (1916); F.W.T.A., ed. 2, 1: 611 (1958), *non* Hassk., *nom illegit.*
F. kisantuensis Warb. in Ann. Mus. Congo, Bot., sér. 6, 1: 22, t. 5 (1904); I.T.U., ed. 2: 252 (1952). Type: Zaire, Kisantu, *Gillet* 598 (B, holo.!, BR, iso.!)

NOTE. The material from Uganda differs from the W. African material in the narrower, thinner, and very shortly acuminate leaves.

36. F. chirindensis *C.C. Berg* in K.B. 43: 78, fig. 1 (1988). Type: Zimbabwe, Chipinga District, Chirinda Forest, *Goldsmith* 23/64 (SRGH, holo.!, BR, K, WAG, iso.!)

Tree up to 35 m. tall, often (?) with pillar-roots. Leafy twigs 1.5–3 mm. thick, minutely puberulous, mostly dark brown to blackish when dry. Leaves in spirals; lamina ± coriaceous, oblong to elliptic or ± ovate, (4–)6–12(–16) × (2.5–)3–5.5(–7.5) cm., apex ± acuminate, base cordate to emarginate or truncate to rarely rounded or obtuse, margin entire; upper surface minutely puberulous on the midrib or glabrous, lower surface minutely puberulous on the main veins, sometimes glabrous; lateral veins (6–)8–12 pairs; tertiary venation reticulate; petiole (1.5–)2–4(–6) cm. long, 1.5–2 mm. thick; stipules 0.3–0.5 cm. long, up to 4 cm. on flush, with appressed (on long stipules ± patent) white hairs outside, caducous. Figs up to 3 together on spurs up to 2 cm. long on the older wood, down to the main branches; bud-scales on the spurs glabrous or sparsely puberulous; peduncle (1–)1.5–2(–4) cm. long, ± 1.5 mm. thick; basal bracts ± 2.5(–3) mm. long, caducous. Receptacle subglobose, 2.5–4 cm. in diameter when fresh, 1.5–3 cm. when dry, minutely puberulous, greenish to pale yellow at maturity, with brown spots; wall ±1 mm. thick when dry.

KENYA. Fort Hall District: Thika R., Thika, near Bulley's Tannery, 24 Mar. 1967, *Faden* 67/149!
TANZANIA. Moshi District: Kibosho, July 1930, *Doughty* 16!; Rungwe District: Kyimbila, 1913, *Stolz* 1907! & 8 Jan. 1914, *Stolz* 2415!
DISTR. **K** 4; **T** 2, 7; Zaire, Malawi, Mozambique and Zimbabwe
HAB. Evergreen forest, riverine; 1500–1700 m.

37. F. sansibarica *Warb.* in E.J. 20: 171 (1894); Hutch. in F.T.A. 6(2): 130 (1916); Peter, F.D.O.-A. 2: 99 (1932); T.T.C.L.: 356 (1949); U.O.P.Z.: 264 (1949). Type: Zanzibar I., *Stuhlmann* 793 (B, holo.!)

Tree up to 20(–40) m. tall, hemi-epiphytic (and strangling) or (secondarily) terrestrial. Leafy twigs 2–5 mm. thick, glabrous or sparsely minutely puberulous, periderm often flaking off older parts at least when dry. Leaves in spirals; lamina coriaceous or subcoriaceous, at least the midrib beneath (and the petiole) usually drying reddish brown, oblong to lanceolate, elliptic or ovate, 4.5–13(–24) × 2–6(–11.5) cm., apex acuminate to subacute or obtuse to occasionally rounded, base rounded to cordate or subacute, margin entire; both surfaces glabrous; lateral veins 5–10(–14) pairs; tertiary venation predominantly reticulate; petiole (0.8–)2–5.5(–8) cm. long, 1–2(–3) mm. thick; stipules 0.1–1.5 cm. long, up to 4.5 cm. on flush, sparsely to densely puberulous or only ciliolate, caducous or on the flush subpersistent. Figs 2–4 together on short (up to 2(–5) cm. long) branched and finally ± cushion-shaped or on up to 15 cm. long straight, or peg-like or sometimes curved spurs on the main or also the lesser branches; bud-scales on the spurs densely puberulous; peduncle 1.2–2.5(–5) cm. long, 2–3 mm. thick; basal bracts 3–5 mm. long, free parts caducous, sometimes subpersistent. Receptacle when dry often stipitate, subglobose, 2–6(–10) cm. in diameter when fresh, 1.5–3(–6) cm. when dry, puberulous, greenish or partly purplish at maturity; wall 5–10 mm. thick when fresh, 2–4(–5) mm. and ± wrinkled when dry.

subsp. **sansibarica**

Lamina generally oblong to lanceolate, often less than 10 cm. long; lateral veins 5–10(–14) pairs; stipules ciliolate. Fig-bearing spurs up to 3.5(–5) cm. long.

KENYA. Kilifi District: Kikambala, Shimo la Tewa, 25 Nov. 1971, *Bally & Smith* 14397! & Kilifi, Oct. 1962, *Dale* 2008! & near Kaloleni, 1 Sept. 1959, *Verdcourt* 2409!
TANZANIA. Pangani, 26 Dec. 1956, *Verdcourt* 1739!; Buha District: Gombe Stream National Park, Kakombe valley, 24 Aug. 1969, *Clutton-Brock* 252!; Kilwa District: Selous Game Reserve, Libungani, 22 Feb. 1971, *Ludanga* 1259!; Zanzibar I., Marahubi, 10 Jan. 1930, *Vaughan* 1094!

DISTR. **K** 7; **T** 3, 4, 6, 7, 8; **Z**; Mozambique, S. Zambia, Zimbabwe and South Africa (Transvaal, Natal)
HAB. Evergreen forest, including drier types, and coastal bushland; 0–900 m.

SYN. *F. langenburgii* Warb. in E.J. 30: 293 (1901). Type: Tanzania, Njombe District, Lumbira [Langenburg], near mouth of Rumbira R., *Goetze* 859 (B, holo.!, K, iso.!)
F. delagoensis Sim, For. Fl. Port. E. Afr.: 99, t. 92 (1909). Type: Mozambique, Maputo, Delagoa Bay, *Sim* 5171 (K, holo.!)

subsp. **macrosperma** (*Mildbr. & Burret*) *C.C. Berg* in K.B. 43: 94 (1988). Lectotype: Cameroun, Bipinde, *Zenker* 2639 (B, lecto., BM, K, isolecto.!)

Lamina oblong to lanceolate and often less than 10 cm. long, or elliptic to oblong to ± ovate and then often over 10 cm. long; lateral veins 8–12 pairs; stipules glabrous. Fig-bearing spurs up to 10(–15) cm. long.

UGANDA. Bunyoro District: Budongo Forest, Sonso R., Nov. 1935, *Eggeling* 2299! & Budongo Forest, Aug. 1935, *Eggeling* 1747!; Masaka District: Sese Is., Sozi, Dec. 1922, *Maitland* 430!
DISTR. **U** 2, 4; west to Sierra Leone, Angola and N. Zambia
HAB. Rain-forest, lakesides and riverine; 1050–1200 m.

SYN. *F. brachylepis* Hiern, Cat. Afr. Pl. Welw. 4: 1011 (1900); Hutch. in F.T.A. 6(2): 126 (1916); I.T.U., ed. 2: 242 (1952); F.F.N.R.: 32 (1962); Hamilton, Ug. For. Trees: 100 (1981); Troupin, Fl. Pl. Lign. Rwanda: 440, fig. 149.2 (1982). Type: Angola, Golungo Alto, *Welwitsch* 6338 (BM, lecto.!)
F. macrosperma Mildbr. & Burret in E.J. 46: 223 (1911); Hutch. in F.T.A. 6(2): 130 (1916); F.W.T.A., ed. 2, 1: 611 (1958); C.C. Berg et al. in Fl. Cameroun 28: 220, t. 78/5 (1985)
F. ugandensis Hutch. in K.B. 1915: 321 (1915) & in F.T.A. 6(2): 129 (1916). Type: Uganda, Masaka District, Buddu, *Dawe* 256 (K, holo.!)
F. gossweileri Hutch. in K.B. 1915: 321, fig. (1915). Type: Angola, Malange, M'Bango Woods, *Gossweiler* 1005 (K, holo.!)

NOTE. The type of *F. ugandensis* (like the type of *F. brachylepis*) deviates somewhat (e.g. in the colour and texture of the dried leaves) from most collections of subsp. *macrosperma*. The differences may be ascribed to the young state of the leaves.

38. F. polita *Vahl*, Enum. Pl. 2: 182 (1805); Hutch. in F.T.A. 6(2): 124 (1916); Peter, F.D.O.-A. 2: 99 (1932); Lebrun & Boutique in F.C.B. 1: 135, t. 14 (1948); T.T.C.L.: 356 (1949); I.T.U., ed. 2: 257 (1952); F.P.S. 2: 269 (1952); F.W.T.A., ed. 2, 1: 611 (1958); Hamilton, Ug. For. Trees: 100 (1981); C.C. Berg in Fl. Cameroun 28: 227, t. 81 (1985). Type: Ghana, *Isert* (C, holo.!, B, iso.!)

Tree up to 15(–40) m. tall, hemi-epiphytic or (secondarily) terrestrial. Leafy twigs 2–5 mm. thick, glabrous or minutely yellowish puberulous, periderm not flaking off. Leaves in spirals; lamina coriaceous or subcoriaceous, at least the midrib beneath (and the petiole) often drying blackish, ovate to oblong or almost elliptic, 5–16(–24) × 3.5–10(–15) cm., apex acuminate, base cordate to truncate or rounded, sometimes subacute, margin entire; both surfaces glabrous; lateral veins 5–12 pairs, tertiary venation partly scalariform to reticulate; petiole 2–12 cm. long, 1–2 mm. thick; stipules 0.5–2 cm. long, glabrous, caducous. Figs 1–4 together on spurs up to 3 cm. long on the older wood; bud-scales of the spurs glabrous; peduncle 0.8–2 cm. long; basal bracts 3–5 mm. long, persistent. Receptacle globose to obovoid, often shortly stipitate at least when dry, 2–4 cm. in diameter when fresh, 1.5–4 cm. when dry, whitish puberulous, greenish to purplish at maturity, wall 2–3 mm. thick and wrinkled when dry.

subsp. **polita**

Lamina ovate to elliptic; lateral veins 5–8(–9) pairs. Peduncle 1–2 cm. long. Receptacle (2–)3–4 cm. in diameter when fresh, 2–4 cm. when dry.

UGANDA. Bunyoro District: Budongo Forest, Apr. 1935, *Eggeling* 1563!; Mengo District: Mabira Forest, Mulange, Sept. 1919, *Dummer* 4307! & Kajansi Forest, Feb. 1938, *Chandler* 2167!
KENYA. Machakos District: Kibwezi, 15 Apr. 1960, *Verdcourt & Polhill* 2689!
TANZANIA. Rufiji District: Mafia I., Kikuni, 13 Aug. 1937, *Greenway* 5082!; Zanzibar I., Prison I., 9 Feb. 1929, *Greenway* 1389! & Bweleo, 18 Oct. 1930, *Vaughan* 1663!
DISTR. **U** 2, 4; **K** 4; **T** 6, 7; **Z**; extending to Senegal, Angola, South Africa (Natal) and Madagascar
HAB. In rain-forest in Uganda, also coastal and lowland bushland in Kenya and Tanzania; 0–1200 m.

subsp. **brevipedunculata** *C.C. Berg* in K.B. 43: 93 (1988). Type: Malawi, N. Province, Misuku Forest, *Chapman* 254 (FHO, holo.!, K, PRE, iso.!)

Lamina oblong to ovate; lateral veins (8–)10–12 pairs. Peduncle 0.8–1.2 cm. long. Receptacle 2–2.5 cm. in diameter when fresh, 1.5–2 cm. when dry.

TANZANIA. Morogoro District, without precise locality, 23 Nov. 1932, *Wallace* 477!
DISTR. T 6; Malawi, Zambia
HAB. Not recorded, but elsewhere in upland evergreen forest; above 1500 m.

39. F. bubu *Warb.* in Ann. Mus. Congo, Bot.,sér. 6, 1: 3, t. 8 (1904); Hutch. in F.T.A. 6(2): 166 (1916); Peter, F.D.O.-A. 2: 102 (1932); Lebrun & Boutique in F.C.B. 1: 160 (1948); T.T.C.L.: 358 (1949); C.C. Berg et al. in Fl. Cameroun 28: 228, t. 82 (1985). Type: Zaire, Kisantu, *Gillet* 1167 (BR, holo.!)

Tree up to 20(–30) m. tall, hemi-epiphytic, often terrestrial; bark pale green to whitish. Leafy twigs 6–12 mm. thick, glabrous or minutely puberulous, periderm often flaking off when dry. Leaves in spirals; lamina coriaceous, elliptic or sometimes oblong to subcircular, 12–30 × 6–23 cm., apex shortly acuminate to almost rounded, base obtuse to rounded or (especially in large leaves) to cordate, margin entire: both surfaces glabrous; lateral veins 6–8(–9) pairs, often divided far from the margin; tertiary venation partly scalariform; petiole 3.5–11(–16) cm. long, 2–5 mm. thick; stipules 0.3–0.5 cm. long, up to 4 cm. on flush, glabrous or partly puberulous, caducous. Figs on short (often almost spine-like) spurs on the main branches (or the trunk); bud-scales of the spurs minutely puberulous, mostly apiculate; peduncle 0.7–1 cm. long, 2–2.5 mm. thick; basal bracts 4–5 mm. long, persistent. Receptacle globose, ± 3 cm. in diameter when fresh, ± 2.5 cm. when dry, glabrous or minutely puberulous, brownish at maturity; wall wrinkled when dry.

UGANDA. Bunyoro District: Budongo Forest, 14 Oct. 1962, *Reynolds* 8!; Toro District: Bwamba, Semliki Forest, Feb. 1943, *St. Clair Thompson* in *Eggeling* 5227!; Mengo District: Kajansi Forest, July 1937, *Chandler* 1783!
KENYA. Kilifi District: Kakoneni, Nov. 1962, *Dale* 2014!
TANZANIA. Lushoto District: Mombo Forest Reserve, 25 Nov. 1966, *Semsei* 4111!; Tanga District: Magunga Estate, 22 July 1954, *Faulkner* 1471!; Kilwa District: Kingupira Forest, 7 Sept. 1977, *Vollesen* in *MRC* 4679!
DISTR. **U** 2, 4; **K** 7; **T** 3, 5, 7, 8; extending to South Africa, W. Angola and Ivory Coast (or ? Senegal)
HAB. Forest, riverine, lakesides and other ground-water forest, sometimes persisting in disturbed places; 0–1200 m.

SYN. *F. kyimbilensis* Mildbr. in Willdenowia 1: 27 (1953). Type: Tanzania, Rungwe District, Masoko [Massoko], Mbaka R., *Stolz* 1827 (B, holo.!, K, iso.!)

40. F. ovata *Vahl*, Enum. Pl. 2: 185 (1805); Hutch. in F.T.A. 6(2): 164 (1916); Peter, F.D.O.-A. 2: 101 (1932); Lebrun & Boutique in F.C.B. 1: 160 (1948); T.T.C.L.: 358 (1949); F.P.S. 2: 270 (1952); F.W.T.A., ed. 2, 1: 608 (1958); Troupin, Fl. Pl. Lign. Rwanda: 448, fig. 150.2 (1982); C.C. Berg et al. in Fl. Cameroun 28: 230, t. 83 (1985). Type: Ghana, *Thonning* 246 (S, holo.)

Tree up to 10(–25) m. tall, hemi-epiphytic (and strangling) or terrestrial, sometimes a shrub or lianescent. Leafy twigs 6–12 mm. thick, densely white to yellowish puberulous or pubescent to almost glabrous, periderm not flaking off. Leaves in spirals; lamina coriaceous, ovate to elliptic, subovate or oblong, (5–)9–31 × (3.5–)6–20 cm., apex acuminate, base cordate to truncate, obtuse or subacute, margin entire; upper surface glabrous or sparsely puberulous in the lower part of the lamina, lower surface sparsely to densely white (or brownish) puberulous to hirtellous or glabrous; lateral veins 10–14 pairs, the basal pair ± faintly branched, reaching the margin far below the middle of the lamina; tertiary venation partly scalariform; petiole 3–10(–13) cm. long, 2–4 mm. thick; stipules 0.3–1 cm. long, glabrous or puberulous to pubescent, caducous. Figs solitary (or in pairs) in the leaf-axils, just below the leaves, or sometimes also on the older wood, initially enveloped by an ovoid calyptrate coriaceous bud-cover up to 2.5 cm. long; peduncle up to 0.5(–1) cm. long, (3–)4–6 mm. thick; basal bracts ± 3–4 mm. long, persistent. Receptacle ovoid to ellipsoid, sometimes subglobose or obovoid, 3–5 × 2.5–4.5 cm. in diameter when fresh, 1–4 × 1–3 cm. when dry, puberulous to pubescent, greenish at maturity.

UGANDA. W. Nile District: War [Warr], Mar. 1935, *Eggeling* 1881!; Kigezi District: Kamwezi, July 1949, *Purseglove* 2993!; Mengo District: Kampala, Makerere College, 22 Feb. 1961, *A.K. Miller* 377!
KENYA. N. Kavirondo District: Malikisi [Malakisi], Feb. 1960, *Templer* T.9; S. Kavirondo District: Watende, 14 Apr. 1955, *Argyle* 112!; Masai District: Mara Plains, Keekorok [Egalok], 21 Oct. 1958, *Verdcourt* 2291A!
TANZANIA. Musoma District: Mara R., Wogakuria [Wodakurea], 19 Feb. 1968, *Greenway et al.* 13305!; Kigoma District: Kanangiye, 27 Mar. 1964, *Pirozynski* 607!; Mpanda District: Mahali Mts., Kasoje, 25 Sept. 1958, *Newbould & Jefford* 2634!

DISTR. **U** 1–4; **K** 3, 5, 6; **T** 1, 4, 7; extending to Senegal and Ethiopia, also to Mozambique, Malawi, N. Zambia and N. Angola
HAB. Deciduous woodland and wooded grassland, riverine, lakesides, often planted in villages; 750–2100 m.

SYN. *Urostigma ovatum* (Vahl) Miq. in Hook., Lond. Journ. Bot. 6: 553 (1847)
Ficus octomelifolia Warb. in Ann. Mus. Congo, Bot., sér 6, 1: 1 (1904). Type: Zaire, without locality, *Cabra* (BR, holo.!)
F. ovata Vahl. var. *octomelifolia* (Warb.) Mildbr. & Burret in E.J. 46: 244 (1911); Peter, F.D.O.-A. 2: 101 (1932); Lebrun & Boutique in F.C.B. 1: 161 (1948); T.T.C.L.: 358 (1949)
F. brachypoda Hutch. in K.B. 1915: 339 (1915) & in F.T.A. 6(2): 189 (1916); Peter, F.D.O.-A. 2: 103 (1932); T.T.C.L.: 360 (1949); I.T.U., ed. 2: 242 (1952); K.T.S.: 315 (1961); F.F.N.R.: 33 (1962); Hamilton, Ug. For. Trees: 99 (1981); Troupin, Fl. Pl. Lign. Rwanda: 440, 149.1 (1982), *non* Miq. (1847), *nom. illegit.* Type: Uganda, Masaka District, Buddu, *Dawe* 290 (K, holo.!)

41. F. pseudomangifera *Hutch.* in K.B. 1915: 342 (1915) & in F.T.A. 6(2): 204 (1917); Peter, F.D.O.-A. 2: 107 (1932); I.T.U., ed. 2; 257 (1952); F.W.T.A., ed. 2, 1: 610 (1958); Hamilton, Ug. For. Trees: 104 (1981); Troupin, Fl. Pl. Lign. Rwanda: 448 (1982); C.C. Berg et al. in Fl. Cameroun 28: 192, t. 66 (1985). Type: Zaire, Lake Kivu, Wau I., *Mildbraed* 1145 (K, lecto.!)

Tree up to 10 m. tall, hemi-epiphytic. Leafy twigs 3–6 mm. thick, glabrous or minutely brownish puberulous, periderm not flaking off. Leaves in spirals or tending to distichous; lamina coriaceous, oblong to lanceolate, sometimes elliptic, 8–32 × 2–9 cm., apex acuminate, sometimes sharply so, base rounded to ± cordate or acute, margin entire; both surfaces glabrous; lateral veins 14–27 pairs, tertiary venation parallel to the lateral veins or reticulate; petiole 1–3 cm. long, ± 3 mm. thick; stipules 2–5 mm. long, ± densely greyish to yellowish or brownish strigose to subsericeous. Figs up to 6 together on short spurs in the leaf-axils or just below the leaves; peduncle 0.3–0.8 cm. long; basal bracts 2–3 mm. long, persistent. Receptacle ± globose, 0.6–1.2 cm. in diameter when fresh, 0.3–0.7 cm. when dry, minutely puberulous, orange to red at maturity.

UGANDA. Bunyoro District: Budongo Forest, Nov. 1935, *Eggeling* 2263!; Masaka District: Nkose I., 21 Jan. 1956, *Dawkins* 885!
DISTR. **U** 2, 4; extending to Sierra Leone
HAB. Rain-forest; 1050–1200 m.

SYN. *F. mangiferoides* Hutch. in K.B. 1915: 342 (1915) & in F.T.A. 6(2): 205 (1917); Lebrun & Boutique in F.C.B. 1: 166 (1948); F.W.T.A., ed. 2, 1: 611 (1958). Type: Cameroun, Bipinde, *Zenker* 1690 (K, lecto.!, B, BR, L, P, isolecto.!)

NOTE. Also planted in the National Museums grounds on Museum Hill, Nairobi, e.g. *Fosberg* 49859! & *Verdcourt* 1028!

42. F. usambarensis *Warb.* in E.J. 20: 159 (1894); P.O.A. C: 162, 11A–E (1895); Hutch. in F.T.A. 6(2): 135 (1916); Peter, F.D.O.-A. 2: 100 (1932); T.T.C.L.: 357 (1949). Type: Tanzania, Tanga District, Amboni, *Holst* 2897 (B, holo.!, K, iso.!)

Tree up to 15 m. tall (? secondarily) terrestrial. Leafy twigs 5–7 mm. thick, puberulous to almost glabrous, periderm not flaking off. Leaves in spirals; lamina coriaceous, oblong to lanceolate, 9.5–14.5 × 4.5–6 cm., apex obtuse to rounded, base obtuse to rounded, margin entire; both surfaces glabrous; lateral veins 10–13 pairs; tertiary venation largely parallel to the lateral veins; petiole 1.5–3.5 cm. long, 2–2.5 mm. thick; stipules 0.3–0.5 cm. long, white to yellowish puberulous. Figs 2–7 together on small spurs in the leaf-axils or just below the leaves; peduncle 0.8–1.7 cm. long; basal bracts ± 3 mm. long, persistent. Receptacle globose, 0.8–1 cm. in diameter when dry, sparsely puberulous.

TANZANIA. Tanga District: Pongwe, 13 Jan. 1931, *Greenway* 4844! & Amboni, 18 July 1932, *Geilinger* 979!; Buha District: Gombe Stream Reserve, Kakombe, 24 Aug. 1973, *Wrangham* G.7111!
DISTR. **T** 3, 4; not known elsewhere
HAB. Woodland near coast, with *Adansonia*, *Pteleopsis* and *Tetracera*, and along coast of Lake Tanganyika with *Anthocleista*; 50–1000 m.

43. F. oreodryadum *Mildbr.* in E.J. 46: 240 (1911); Hutch. in F.T.A. 6(2): 187 (1916); Peter, F.D.O.-A. 2: 109 (1932); Lebrun & Boutique in F.C.B. 1: 147 (1948); Troupin, Fl. Pl. Lign. Rwanda: 466, fig. 147.4 (1982). Type: Rwanda, Rukarara-Rugege, *Mildbraed* 1031 (B, holo.!, K, iso.!)

Large tree, or shrub, hemi-epiphytic (and strangling). Leafy twigs 2.5–6 mm. thick, almost glabrous, periderm not flaking off. Leaves in spirals; lamina coriaceous, oblong to elliptic or subovate, 9–18 × 3.5–6 cm., apex acuminate, base acute to obtuse, sometimes subcordate, margin entire; both surfaces glabrous; lateral veins 8–14 pairs, tertiary venation reticulate; petiole 2–5 cm. long, 1.5–2 mm. thick; stipules ± 1.5 cm. long, yellowish appressed puberulous or glabrous, caducous. Figs in pairs in the leaf-axils, sometimes also just below the leaves, sessile; basal bracts ± 4 mm. long, persistent. Receptacle subglobose to ellipsoid or ovoid, ± 2 cm. in diameter when fresh, ± 1.5 cm. when dry, almost glabrous, ± warted, orange to yellow at maturity.

UGANDA. Kigezi District: Rutenga, Oct. 1940, *Eggeling* 4211!
DISTR. U 2; Rwanda, Burundi, E. Zaire, Cameroun and Bioko [Fernando Po]
HAB. Recorded in Uganda only from swamp edge, but elsewhere in montane and submontane rain-forest between 1300 and 2500 m.

44. F. barteri *Sprague* in Gard. Chron., ser. 3, 33: 354 (1903); Hutch. in F.T.A. 6(2): 205 (1917); Lebrun & Boutique in F.C.B. 1: 164 (1948); F.W.T.A., ed. 2, 1: 610 (1958); F.F.N.R.: 33 (1962); C.C. Berg et al. in Fl. Cameroun 28: 190, t. 65 (1985). Type: Nigeria, Onitsa, *Barter* 294 (K, holo.!, BR, P, iso.!)

Tree up to 10 m. tall or a shrub, hemi-epiphytic. Leafy twigs 3–5 mm. thick, glabrous, periderm not flaking off. Leaves in spirals; lamina coriaceous, lanceolate to linear, or less often oblong to elliptic, (5.5–)10–18(–30) × 1.5–3.5(–7) cm., apex acuminate to subacute, base acute or less often rounded, margin entire; both surfaces glabrous; lateral veins 10–20 pairs, tertiary venation reticulate or parallel to the lateral veins; petiole 1–4 cm. long, ± 2 mm. thick; stipules 0.5–2 cm. long, glabrous, caducous. Figs in pairs in the leaf-axils; peduncle (0.5–)1–2.5 cm. long; basal bracts 1.5–2 mm. long, caducous. Receptacle globose, 1–1.5 cm. in diameter when fresh, ± 0.5–1 cm. when dry, glabrous, smooth to warted, yellow to orange at maturity.

UGANDA. Bunyoro District: Budongo Forest, Nov. 1938, *Eggeling* 3452!; Mengo District: Nagojje [Nagoje], Jan. 1918, *Dummer* 3289!
DISTR. U 2, 4; extending to N. Zambia and to Sierra Leone
HAB. In rain-forest and drier types of evergreen forest, altitude not recorded, ± 1000–1200 m.

SYN. [*F. stipulifera* sensu I.T.U., ed. 2: 258 (1952), pro parte, *non* Hutch.]

NOTE. The species is very variable in the shape of the lamina. In the material from Uganda the narrow-leaved form is represented.

45. F. conraui *Warb.* in Ann. Mus. Congo., Bot., sér. 6, 1: 25, t. 11 (1904); Hutch. in F.T.A. 6(2): 150 (1916); Lebrun & Boutique in F.C.B. 1: 155 (1948); F.W.T.A., ed. 2, 1: 607 (1958); C.C. Berg et al. in Fl. Cameroun 28: 234, t. 84 (1985). Type: Cameroun, Bangwe, *Conrau* 280 (B, holo.!, Z, iso.!)

Treelet or shrub, sometimes lianescent, hemi-epiphytic. Leafy twigs 2–4 mm. thick, glabrous or puberulous, periderm not flaking off. Leaves in spirals; lamina coriaceous, oblong to lanceolate, subobovate or elliptic, (6–)10–20(–28) × (2–)3.5–7 cm., apex acuminate, base acute to rounded, margin entire; both surfaces glabrous; lateral veins 8–11 pairs, the lower ones relatively small, tertiary venation reticulate; petiole 0.8–4.5 cm. long, 1–2 mm. thick, epiderm flaking off when dry or not; stipules (0.5–)1–2.5 cm. long, basally connate, glabrous, ± persistent. Figs in pairs or solitary in the leaf-axils, sessile; basal bracts 2–4 mm. long. persistent. Receptacle subglobose to obovoid or sometimes subpyriform, 2–3 cm. in diameter when fresh, (0.5–)1.5–2.5 cm. when dry, glabrous or subhispid, verrucose or smooth, at maturity greenish with reddish spots or warts or partly reddish; wall 1–4 mm. thick when dry.

UGANDA. Masaka District: S. Buddu, *Dawe* 301!
DISTR. U 4; extending to Sierra Leone, Gabon and Angola
HAB. Not recorded in Uganda, but elsewhere in areas with rain-forest, drier types of evergreen or semi-deciduous forest, at altitudes up to 1200(–1500) m.

SYN. *F. stipulifera* Hutch. in K.B. 1915: 326 (1915); I.T.U., ed. 2: 258, fig. 58b (1952), pro parte. Type: Uganda, Masaka District, S. Buddu, *Dawe* 301 (K, holo.!)

46. F. cyathistipula *Warb.* in E.J. 20: 173 (1894); P.O.A. C: 161, fig. 10A–E (1895); Hutch. in F.T.A. 6(2): 153 (1916); Peter in F.D.O.-A. 2: 103 (1932); T.T.C.L.: 357 (1949); Lebrun & Boutique in F.C.B. 1: 172 (1948); I.T.U., ed. 2: 245, fig. 58a (1952); F.F.N.R.: 32, t. 6H (1962);

Hamilton, Ug. For. Trees: 100 (1981); Troupin, Fl. Pl. Lign. Rwanda: 442, fig. 150.4 (1982); C.C. Berg et al. in Fl. Cameroun 28: 246, t. 89 (1985). Type: Tanzania, Bukoba, *Stuhlmann* 3779 (B, lecto.!)

Tree up to 8(–15) m. tall, terrestrial or hemi-epiphytic. Leafy twigs 3–5 mm. thick, glabrous or white puberulous, periderm sometimes flaking off when dry. Leaves in spirals; lamina coriaceous, oblanceolate to obovate, 6–20 × 3–8 cm., apex acuminate, base acute to attenuate, margin entire; both surfaces glabrous; lateral veins 5–7(–8) pairs, tertiary venation reticulate; petiole 1.5–4 cm. long, ± 3 mm. thick, epiderm not flaking off; stipules partly connate, (0.5–)1–2(–3) cm. long, minutely white puberulous or almost glabrous, persistent. Figs 1(–3) in the leaf-axils; peduncle 0.5–2.5 cm. long in E. Africa; basal bracts ± 4 mm. long, persistent. Receptacle globose to obovoid (and usually on stipes up to 1 cm. long) or pyriform, 3–5 cm. in diameter when fresh, 2–3 cm. when dry, often somewhat scabrous, pale green to pale yellow at maturity; wall up to 8 mm. thick and spongy, when dry 2–5 mm. thick, smooth, warted or sometimes with protuberances up to 3 mm. long; apex protruding up to 1 cm. when dry.

UGANDA. Ankole District: Lake Lutoto, Aug. 1936, *Eggeling* 3170!; Busoga District: Dumba ferry, July 1926, *Maitland* 1090!; Mengo District: Entebbe, Botanic Gardens, 7 June 1946, *A.S. Thomas* 4485!

KENYA. Nandi District: Devils Gorge, 15 Dec. 1938, *Green* 12!; Fort Hall District: Thika, 31 Oct. 1966, *Faden* 66/125; N. Kavirondo District: Kakamega Forest, near Forest Station, 4 Jan. 1968, *Perdue & Kibuwa* 9453!

TANZANIA. Lushoto District: Amani, 13 Nov. 1935, *Greenway* 4172!; Tanga District: Potwe Forest, 22 Aug. 1961, *Mgaza* 426!; Morogoro District: Uluguru Mts., Mkungwe Hill, 5 July 1970, *Faden* in *Kabuye* 289!

DISTR. **U** 1–4; **K** 4, 5; **T** 1, 3, 4, 6–8; extending to Ivory Coast, Angola and N. Zambia

HAB. Rain-forest, lakesides, riverine, ground-water forest, sometimes on rocks; 700–1800 m.

SYN. *F. rhynchocarpa* Mildbr. & Burret in E.J. 46: 235 (1911); Hutch. in F.T.A. 6(2): 114 (1916); Peter, F.D.O.-A. 2: 102 (1932); T.T.C.L.: 357 (1949); K.T.S. 319 (1961). Type: Tanzania, Lushoto District, Ngwelo [Nguelo], *Kummer* 25 (B, lecto.!)

F. nyanzensis Hutch. in K.B. 1915: 327 (1915). Type: Uganda, Lake Victoria [Nyanza], *Bagshawe* 690 (K, holo.!, BM, iso.!)

F. rederi Hutch. in K.B. 1915: 329(1915); F.W.T.A., ed. 2, 1: 608 (1958). Type Cameroun, Buea, *Reder* 395 (B, holo.!)

NOTE. Subsp. *pringsheimiana* (Braun & K. Schum.) C.C. Berg, with sessile figs, occurs in Cameroun.

47. F. densistipulata *De Wild.* in F.R. 12: 194 (1913); Hutch. in F.T.A. 6(2): 151 (1916). Type: NE. Zaire, without precise locality, *Seret* 734 (BR, holo.!, P, iso.!)

Tree up to 10 m. tall or a shrub, hemi-epiphytic. Leafy twigs 2.5–5 mm. thick, puberulous to hirtellous, periderm not flaking off. Leaves in spirals; lamina coriaceous, oblong to lanceolate or subobovate, 12–25 × 4–7(–11) cm., apex acuminate, base acute to rounded, margin entire; both surfaces glabrous; lateral veins 6–8 pairs, tertiary venation reticulate; petiole 1–2.5(–5) cm. long, 2–4 mm. thick, epiderm flaking off when dry; stipules 1.5–3 cm. long, partly connate, sparsely puberulous, ± persistent. Figs 1–2 in the leaf axils, subsessile or on peduncles up to 0.4 cm. long; basal bracts 2–3.5 mm. long, persistent. Receptacle subglobose, 1.3–2.2 cm. in diameter when fresh, 1–1.5 cm. when dry, with a stipe 4–9 mm. long, sparsely puberulous to almost glabrous, smooth or warted, at maturity greenish or with yellow spots or partly reddish; wall 0.5–1 mm. thick when dry, not spongy; apex almost plane.

UGANDA. Bunyoro District: Budongo Forest, July 1935, *Eggeling* 2126!; Masaka District: Namalala Forest Reserve, 24 Sept. 1957, *Osmaston* 4200! & Nkose I., 21 Jan. 1956, *Dawkins* 887!

DISTR. **U** 2, 4; extending to Cameroun and Gabon

HAB. Rain-forest and semi-swamp forest; 1100–1260 m.

SYN. *F. namalalensis* Hutch. in K.B. 1915 : 328, fig. (1915); I.T.U., ed. 2: 253 (1952). Type: Uganda, Masaka District, Namalala, *Fyffe* 77 (K, lecto.!)

48. F. scassellatii *Pamp.* in Bull. Soc. Bot. Ital. 1915: 15 (1915); Friis in Nordic Journ. Bot. 5: 331 (1985); C.C. Berg in K.B. 43: 94 (1988). Lectotype — see Friis, l.c.: Somalia (S), without precise locality, *Scassellati* (FT, lecto.)

Tree up to 50 m., hemi-epiphytic, secondarily terrestrial. Leafy twigs 3–8 mm. thick, glabrous or minutely puberulous, periderm flaking off older parts when dry. Leaves in spirals; lamina coriaceous, oblong to oblanceolate, obovate or elliptic, 6–20(–28) × 3–8 cm., apex shortly and bluntly acuminate to rounded, base ± acute, margin entire; both surfaces glabrous; lateral veins 8–18 pairs, gradually becoming stronger towards the apex; tertiary venation reticulate; petiole 0.5–2.5(–3.5) cm. long, 2–3 mm. thick, epiderm occasionally flaking off; stipules 0.3–2 cm. long, free, glabrous or minutely puberulous, caducous. Figs 1–2 in the leaf-axils; peduncle 0.5–1.5 cm. long or figs sessile to subsessile; basal bracts 3–5 mm. long, persistent. Receptacle often shortly stipitate at least when dry, globose to ellipsoid, 3–4.5 cm. in diameter when fresh, 1.2–2(–3) cm. in diameter when dry, ± sparsely minutely puberulous, green at maturity; wall 4–5 mm. thick, not spongy; apex in dry material protruding up to 7 mm.

subsp. **scassellatii**

Lamina oblong to subobovate or elliptic, sometimes oblanceolate, up to 20 × 8 cm. Figs on peduncles 0.5–1.5 cm. long. Receptacle 1.2–2 cm. in diameter when dry and the apex ± strongly protruding.

KENYA. Masai District: Mzima Springs, 17 Jan. 1961, *Greenway* 9759!; Teita District: Taveta, Mar. 1937, *Dale* in *F.D.* 3657!; Kilifi District: Marafa, 19 Nov. 1961, *Polhill & Paulo* 807!

TANZANIA. Lushoto District: Uberi–Monga, 26 Jan. 1939, *Greenway* 5828!; Pangani District: Pangani Falls, 22 Sept. 1957, *Faulkner* 2064!; Njombe District: Madehani, 26 Jan. 1914, *Stolz* 2480!; Pemba I., coast, 14 Oct. 1929, *Burtt Davy* 22575!

DISTR. **K** 1, 4, 6, 7; **T** 2, 3, 6, 7; **Z**; **P**; Somalia (S.), Zaire, Malawi, Zimbabwe

HAB. Rain-forest, lakesides, riverine, ground-water forest and sometimes open places along the coast; 1800 m.

SYN. *F. kirkii* Hutch. in K.B. 1915: 343 (1915) & in F.T.A. 6(2): 209 (1917); T.T.C.L.: 361 (1949); U.O.P.Z.: 263 (1949); K.T.S.: 317 (1961). Type: Tanzania, Zanzibar, *Kirk* (K, lecto.!)

subsp. **thikaensis** *C.C. Berg* in K.B. 43: 95 (1988). Type: Kenya, Fort Hall District, Thika, *Faden* 66/27 (K, holo.!, EA, iso.)

Lamina mostly oblanceolate, up to 28 × 4.5 cm. Figs sessile or subsessile. Receptacle 2–3 cm. in diameter when dry, apex hardly protruding.

KENYA. Fort Hall District: Thika, 23 Mar. 1967, *Faden* 67/139! & Thika Falls, 18 Feb. 1951, *Greenway & Verdcourt* 8493!

DISTR. **K** 4; known only from the Chania gorge near Thika

HAB. Riverine forest; 1350–1450 m.

49. F. preussii *Warb.* in E.J. 20: 156 (1894); Hutch. in F.T.A. 6(2): 152 (1916); Lebrun & Boutique in F.C.B. 1: 162 (1948); F.W.T.A., ed. 2, 1: 607 (1958); C.C. Berg et al. in Fl. Cameroun 28: 242, t. 87 (1985). Type: Cameroun, Barombi, *Preuss* 454 (B, holo.!, BM, K, iso.!)

Tree up to 10 m. tall or a shrub, hemi-epiphytic. Leafy twigs 8–10 mm. thick, ± densely white strigose to puberulous, periderm flaking off when dry. Leaves in spirals; lamina coriaceous, oblong to subovate, sometimes subpandurate, 19–26(–45) × 8–13(–17) cm., apex acuminate, base obtuse to subcordate, sometimes cordate, margin entire; upper surface glabrous, lower surface glabrous or white to brownish hirtellous on the lower part of the midrib; lateral veins 6–8(–9) pairs; tertiary venation ± reticulate; petiole 2.5–8.5 cm. long, ± 4 mm. thick, epiderm flaking off when dry; stipules ± 2 cm. long, free or basally connate, glabrous or sometimes white strigillose, persistent. Figs 1–2 in the leaf-axils, sessile or subsessile; basal bracts ± 5 mm. long, persistent. Receptacle globose, 2.5–5 cm. in diameter when fresh, 2–3 cm. when dry, rather densely yellowish to white strigillose to hirtellous, ± pronouncedly warted, at maturity greenish with yellowish spots; wall ± 3–5 mm. thick when dry; apex protruding up to 5 mm.

UGANDA. Mengo District: Kajansi Forest, 16 km. on Entebbe road, Nov. 1937, *Chandler* 1999!

DISTR. **U** 4; extending to Nigeria

HAB. Rain-forest; 1150–1200 m.

50. F. wildemaniana *De Wild. & Th. Dur.* in Ann. Mus. Congo, Bot., sér. 3, 2: 217 (1901); Hutch. in F.T.A. 6(2): 178 (1916); Lebrun & Boutique in F.C.B. 1: 158, t. 17 (1948); C.C. Berg et al. in Fl. Cameroun 28: 258, t. 96 (1985). Type: Zaire, without precise locality, *Dewèvre* 562 (BR, holo.!, B, iso.!)

Tree or shrub, hemi-epiphytic. Leafy twigs 10–20 mm. thick, glabrous or brownish puberulous, periderm not flaking off. Leaves in spirals; lamina coriaceous, lanceolate to oblong or subobovate, often ± pandurate, (15–)25-60 × 6–25 cm., apex acuminate to subacute, base acute or sometimes subattenuate and cordulate, margin entire; both surfaces glabrous; lateral veins 8–14 pairs, tertiary venation partly scalariform; petiole 1.5–8 cm. long, 4–10 mm. thick, epiderm not flaking off; stipules 0.3–1 cm. long, glabrous, caducous. Figs 1–2 in the leaf-axils, sessile; basal bracts 4–10 mm. long, persistent. Receptacle globose to somewhat depressed-globose, 2.5–4 cm. in diameter when dry, yellow hirtellous to strigillose; wall 1–1.5 mm. thick when dry, wrinkled.

UGANDA. Bunyoro District: Budongo Forest, Nyakatoma, 5 May 1962, *Reynolds* 11!
DISTR. U 2; Zaire, Gabon and Cameroun
HAB. Rain-forest; about 1100 m.

Addenda

Streblus usambarensis (*Engl.*) *C.C. Berg* (*Sloetiopsis usambarensis* Engl.) in Konink. Nederl. Akad. Weten., Ser. C, 91: 357 (1988)).

Since going to press the genus *Sloetiopsis* has been included within *Streblus*; *S. usambarensis* is attributed to sect. *Streblus*.

CECROPIACEAE

C.C. BERG
(University of Bergen)

Trees, often with stilt-roots, dioecious, sap watery, turning black. Leaves in spirals; lamina palmately or radiately incised; stipulate. Staminate inflorescences branched; tepals 2–4; stamens 1 or 3–4. Pistillate inflorescences globose- or clavate-capitate; tepals 2–3; pistil 1; ovary free or basally adnate to the perianth; stigma 1; ovule 1, basally attached. Fruit achene-like or forming a drupaceous whole with the fleshy perianth. Seed large and without endosperm or small and with endosperm.

A family with 6 genera and ± 200 species, all in the tropics; 1 genus (± 20 species) in Asia and Australasia; 3 genera (± 200 species) in the Neotropics and 2 genera (9 species) in Africa.

Cecropia peltata L., Trumpet Wood, has been grown in the Entebbe Botanic Garden, see Dale, Introd. Trees Ug.: 21 (1953).

Lamina basally attached, palmately incised	**1. Myrianthus**
Lamina peltate, radially incised	**2. Musanga**

1. MYRIANTHUS

P. Beauv., Fl. Oware 1: 16 (1805); De Ruiter in B.J.B.B. 46: 472 (1976)

Trees or shrubs, often with stilt-roots. Lamina basally attached, palmately incised in Flora area, elsewhere sometimes entire; stipules fully amplexicaul, connate. Inflorescences bracteate. Staminate inflorescences branched; flowers in spike-like to globose glomerules; tepals 3–4; stamens 3–4. Pistillate inflorescences globose-capitate; perianth 2–3-lobed; stigma tongue-shaped. Fruiting perianth enlarged, fleshy, yellow to orange-red; fruit adnate to the perianth; endocarp woody. Seed large, without endosperm.

The genus is African and comprises 7 species, most of them in West and Central Africa.

Lamina hirtellous to puberulous on the main veins above; staminate inflorescence with spike-like glomerules; pistillate inflorescence with ± 20–50 flowers:	
Lamina on the main veins beneath and the stipules with whitish to pale yellow hairs; glomerules of the staminate inflorescence 0.2–0.4 cm. in diameter; apex of the fruiting perianth flat	1. *M. arboreus*
Lamina on the main veins beneath and the stipules with brown to golden-yellow hairs; glomerules of the staminate inflorescence 0.4–0.6 cm. in diameter; apex of the fruiting perianth conical	2. *M. holstii*
Lamina glabrous above; staminate inflorescence with globose to ellipsoid terminal glomerules; pistillate inflorescence with 8–± 20 flowers	3. *M. preussii*

1. **M. arboreus** *P. Beauv.*, Fl. Oware 1: 16, t. 11, 12 (1805); Engl., E.M. 1: 37, t. 16 (1898); Rendle in F.T.A. 6(2): 231 (1917); Peter, F.D.O.-A. 2: 114 (1932); Hauman in F.C.B. 1: 83 (1948); T.T.C.L.: 363 (1949); I.T.U., ed. 2: 264 (1952); F.P.S. 2: 273 (1952); F.W.T.A., ed. 2, 1: 614 (1958); De Ruiter in B.J.B.B. 46: 478, t. 2, 3 (1976); Hamilton, For. Trees Ug.: 201 (1981); C.C. Berg et al. in Fl. Cameroun 28: 262, figs. 98, 99 (1985). Type: Nigeria, Benin, *P. de Beauvois* (G, holo.!)

Tree up to 10 m. tall or a shrub. Lamina ± 30 × 30 to ± 90 × 90 cm., 5–7-fid to -parted down to the petiole; segments sessile or stoutly petiolulate, margin rather irregularly (often coarsely and doubly) crenate- to serrate-dentate or -denticulate; upper surface hirtellous to puberulous on the main veins, lower surface whitish hirtellous on the main veins; petiole (15–) 25–55 cm. long, whitish hirtellous; stipules (1.5–)3–5 cm. long, whitish to pale yellow subsericeous or subhirsute. Staminate inflorescences ± 5–30 cm. in diameter; peduncle (7–)13–21 cm. long; glomerules of flowers spike-like, 2–4 mm. in diameter. Pistillate inflorescences 2–3.5 cm., in fruit 6–10(–15) cm. in diameter; peduncle 2–6 cm. long; flowers ± 20–50; apex of the perianth almost flat; endocarp-body ± 1.6 × 0.7 cm.

UGANDA. Mengo District: Entebbe, Kitubulu, Oct. 1931, *Eggeling* 237! & Kivuvu and Kirerema, Oct. 1913, *Dummer* 421! & Kajansi Forest, June 1937, *Chandler* 1647!

TANZANIA. Buha District: Kakombe valley, 12 Feb. 1970, *Clutton-Brock* 382!; Mpanda District: Kasoje, 25 Sept. 1958, *Newbould & Jefford* 2638! & Kasiha [Kasieha] R., 20 July 1959, *Newbould & Harley* 4495!

DISTR. U 4; T 4; extending to S. Ethiopia, S. Sudan, N. Angola and to Guinée

HAB. Rain-forest or swamp forest, in clearings and regrowth, riverine or lakesides; 700–1200 m.

2. M. holstii *Engl.*, E.M. 1: 37, t. 16A, 17E (1898); Rendle in F.T.A. 6(2): 237 (1917); Peter, F.D.O.-A. 2: 113 (1932); Hauman in F.C.B. 1: 84 (1948); T.T.C.L.: 363 (1949); I.T.U., ed. 2: 264 (1952); De Ruiter in B.J.B.B. 46: 481, t. 4 (1976); Hamilton, For. Trees Ug.: 201 (1981); Troupin, Fl. Pl. Lign. Rwanda: 453, fig. 151.2 (1982). Type: Tanzania, Lushoto District, Lutindi, *Holst* 3308 (B, lecto.!, K, isolecto.!)

Tree up to 20 m. tall. Lamina ± 25 × 25 to 60 × 60 cm., 3–7(–8)-fid or -parted down to the petiole; segments sessile or stoutly petiolulate, margin rather regularly serrate-dentate to subentire; upper surface hirtellous to puberulous on the main veins, lower surface brownish hirtellous to subsericeous on the main veins; petiole 7–35 cm. long, brownish hirtellous to subsericeous; stipules 1.5–4 cm. long, brown to golden-yellow sericeous. Staminate inflorescences ± 4–15 cm. in diameter; peduncle 3–13.5 cm. long; glomerules of flowers spike-like, 4–6 mm. in diameter. Pistillate inflorescences 1–2 cm., in fruit 5–8 cm. in diameter; peduncle 1–5 cm. long; flowers ± 20–40; apex of the perianth broadly conical; endocarp-body ± 1.2 × 0.8 cm. Fig. 22.

UGANDA. Toro District: Kibale Forest, 30 Jan. 1945, *Greenway & Eggeling* 7055!; Ankole District: Kalinzu Forest, Aug. 1936, *Eggeling* 3204!; Kigezi District: Mushongero–Nyakalembe [Nakalembe], 29 Oct. 1929, *Snowden* 1617!

KENYA. Fort Hall District: 24 km. W. of Fort Hall, *Hutchins*!; Meru District: Mt. Kenya, 20 Aug. 1943, *J. Bally* in *Bally* 3210! & Nyambeni Hills, Stone Bridge, 10 Oct. 1960, *Polhill & Verdcourt* 270!

TANZANIA. Lushoto District: Monga Estate, 27 Oct. 1954, *Bryce* 52!; Kilosa District: Ukaguru Mts., Mandege–Masenge, 4 June 1978, *Thulin & Mhoro* 2929!; Songea District: Liwiri-Kiteza [Luwira-Kiteza] Forest Reserve, 20 Oct. 1956, *Semsei* 2518!

DISTR. **U** 2; **K** 4; **T** 3, 6–8; E. Zaire, Rwanda, Burundi, Mozambique, Malawi, N. Zambia, E. Zimbabwe.

HAB. Rain-forest, sometimes at edges, in regrowth and along rivers; 900–2100 m.

SYN. *M. holstii* Engl. var. *quinquesectus* Engl. in E.J. 30: 295 (1901); Peter, F.D.O.-A. 2: 113 (1932). Type: Tanzania, Rungwe District, Mwakaleli [Muakareri], *Goetze* 1312 (B, holo.!, BM, BR, G, L, iso.!)
M. mildbraedii Peter , F.D.O.-A. 2: 113, Descr.: 10, t. II/11 (1932); T.T.C.L.: 363 (1949). Types: Tanzania, *Peter* 7574, 7722 & 15816 (B, syn., not traced)

NOTE. Hamilton (1981) suggests that *M. holstii* is generally ecologically separated from *M. arboreus*, mostly at higher altitudes and mainly on raised sites rather than in swamps.

3. M. preussii *Engl.* in E.J. 20: 149 (1894) & E.M. 1: 40, t. 17A (1898); Hauman in F.C.B. 1: 86 (1948); F.W.T.A., ed. 2, 1: 616 (1958); De Ruiter in B.J.B.B. 46: 486, t. 5 (1976); C.C. Berg et al. in Fl. Cameroun 28: 266, fig. 100 (1985). Type: Cameroon, Barombi, *Preuss* 478 (B, holo.)

Tree up to 15(–25) m. tall or a shrub. Lamina ± 20 × 20 to 30 × 40 cm., 5–7-parted down to the petiole; segments slenderly petiolulate, margin subentire to serrate-dentate; upper surface glabrous, lower surface sparsely appressed puberulous on the main veins; petiole 10–20 cm. long, sparsely appressed puberulous; stipules 0.5–1.5 cm. long, white to grey sericeous. Staminate inflorescences ± 3–10 cm. in diameter; peduncle 4–13 cm. long; glomerules of flowers globose to ellipsoid, ± 3 mm. in diameter. Pistillate inflorescences 1.5–2.5 cm., in fruit up to 5 cm. in diameter; peduncle 2–4 cm. long; flowers 8–± 20; apex of the perianth narrowly conical to apiculate; endocarp-body ± 1.5 × 1.3 cm.

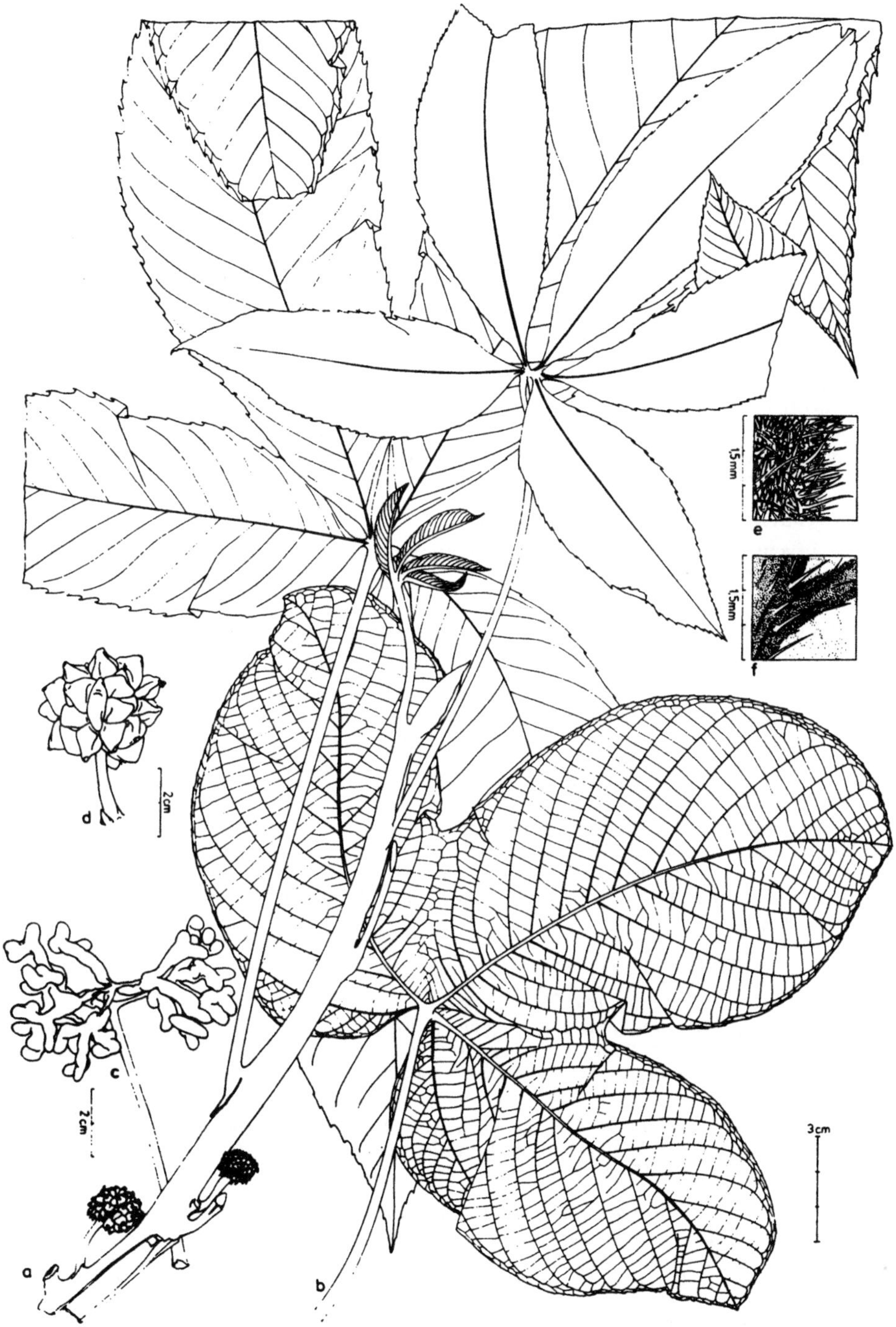

FIG. 22. *MYRIANTHUS HOLSTII* — **1**, leafy twig with pistillate inflorescences; **2**, leaf; **3**, **4**, indumentum of twig and leaf respectively; **5**, staminate inflorescence; **6**, infructescence. 1, 3, from *Stolz* 1591; 2, 4, from *Fries* 2112; 5, from *Mendonça* 295; 6, from *Torre & Correia* 14829. Reproduced with permission from Bulletin du Jardin Botanique National de Belgique.

subsp. **seretii** (*De Wild.*) *De Ruiter* in B.J.B.B. 46: 486, t. 5 (1976). Type: Zaire, Gombari, *Seret* 590 (BR, holo.!)

Shrub or tree 2–10 m. tall. Stipules caducous. Fruiting perianth not protracted, distinctly ovoid; interfloral bracts not lengthened in the infructescence.

TANZANIA. Ulanga District: Sumbagulo R., 15 Sept. 1952, *Carmichael* 117!
DISTR. **T** 6; Zaire and Rwanda
HAB. Riverine forest;

SYN. *M. seretii* De Wild. in Ann. Mus. Congo, Bot., sér. 5, 3: 68 (1909)

2. MUSANGA

R. Br. in Tuckey, Narr. Exped. R. Zaire, App.: 453 (1818); De Ruiter in B.J.B.B. 46: 496 (1976)

Trees with stilt-roots. Lamina peltate, radially incised; stipules fully amplexicaul, connate. Inflorescences bracteate. Staminate inflorescences branched; flowers in globose glomerules; perianth 2-lobed; stamen 1. Pistillate inflorescences clavate-capitate, somewhat compressed; perianth tubular; stigma subpeltate. Fruiting perianth slightly enlarged and fleshy, greenish; fruit free, endocarp crustaceous; mesocarp mucilaginous; exocarp membranaceous. Seed small, with endosperm.

The genus is African and comprises 2 species.

NOTE. In young plants the lamina changes from entire to palmately incised to radiately incised and from basally attached to peltate.

Lamina usually smooth above; segments (9–)11–18 . . . 1. *M. cecropioides*
Lamina scabrous above; segments 8–11 . . . . 2. *M. leo-errerae*

1. M. cecropioides *Tedlie* in Bowdich, Miss. Ashantee: 372 (1819); I.T.U., ed. 2: 263, photo. 44 (1952), pro parte; F.W.T.A., ed. 2, 1: 616 (1958); De Ruiter in B.J.B.B. 46: 500 (1976); Hamilton, For. Trees Ug.: 201 (1981); C.C. Berg et al. in Fl. Cameroun 28: 272, fig. 103 (1985). Type: Zaire, *Chr. Smith* (BM, holo.!, K, iso.!)

Tree up to 30 m. tall. Lamina up to 110 cm. in diameter, with (9–)11–18 segments; upper surface usually smooth, lower surface white arachnoid-tomentellous in the areoles; petiole up to 110 cm. long; stipules 7–30 cm. long. Staminate inflorescences richly branched, glomerules numerous, more than 50. Pistillate inflorescences 2–5(–12 in fruit) cm. long. Fig. 23.

UGANDA. Toro District: Semliki Forest, *Dawe* 642!
DISTR. **U** 2; extending to N. Angola and to Senegal
HAB. Secondary and swamp forest; 750–900 m.

SYN. *M. smithii* Bennett in Bennett & Brown, Pl. Jav. Rar.: 49 (1838); Engl., E.M. 1: 42, t. 18 (1898); Rendle in F.T.A. 6(2): 239 (1917); Peter, F.D.O.-A. 2: 114 (1932); I.T.U.: 147 (1940); Hauman in F.C.B. 1: 88 (1948), *nom. superfl.* Type: as for species

2. M. leo-errerae *Hauman & J. Léon.* in Bull. Agric. Congo Belge 51: 61 (1960); De Ruiter in B.J.B.B. 46: 501 (1976); Hamilton, For. Trees Ug.: 202 (1981); Troupin, Fl. Pl. Lign. Rwanda: 452, fig. 151.1 (1982). Type: Zaire, Bitale, *Pierlot* 1349 (BR, holo.!)

Tree up to 30 m. tall. Lamina up to 40 cm. in diameter, with 8–11 segments; upper surface scabrous, lower surface grey arachnoid-tomentellous in the areoles; petiole 12–35 cm. long; stipules 8–16 cm. long. Staminate inflorescences poorly branched; glomerules 10–25, ± 10 mm. in diameter. Pistillate inflorescences 1.5–2.5(–3 in fruit) cm. long.

UGANDA. Ankole District: Kalinzu Forest, Aug. 1936, *Eggeling* 3206! & June 1938, *Eggeling* 3715!
DISTR. **U** 2; E. Zaire, Rwanda, Burundi
HAB. Forest regrowth; 1350 m.

SYN. [*M. cecropioides* sensu I.T.U., ed. 2: 263 (1952), pro parte, *non* R. Br.]

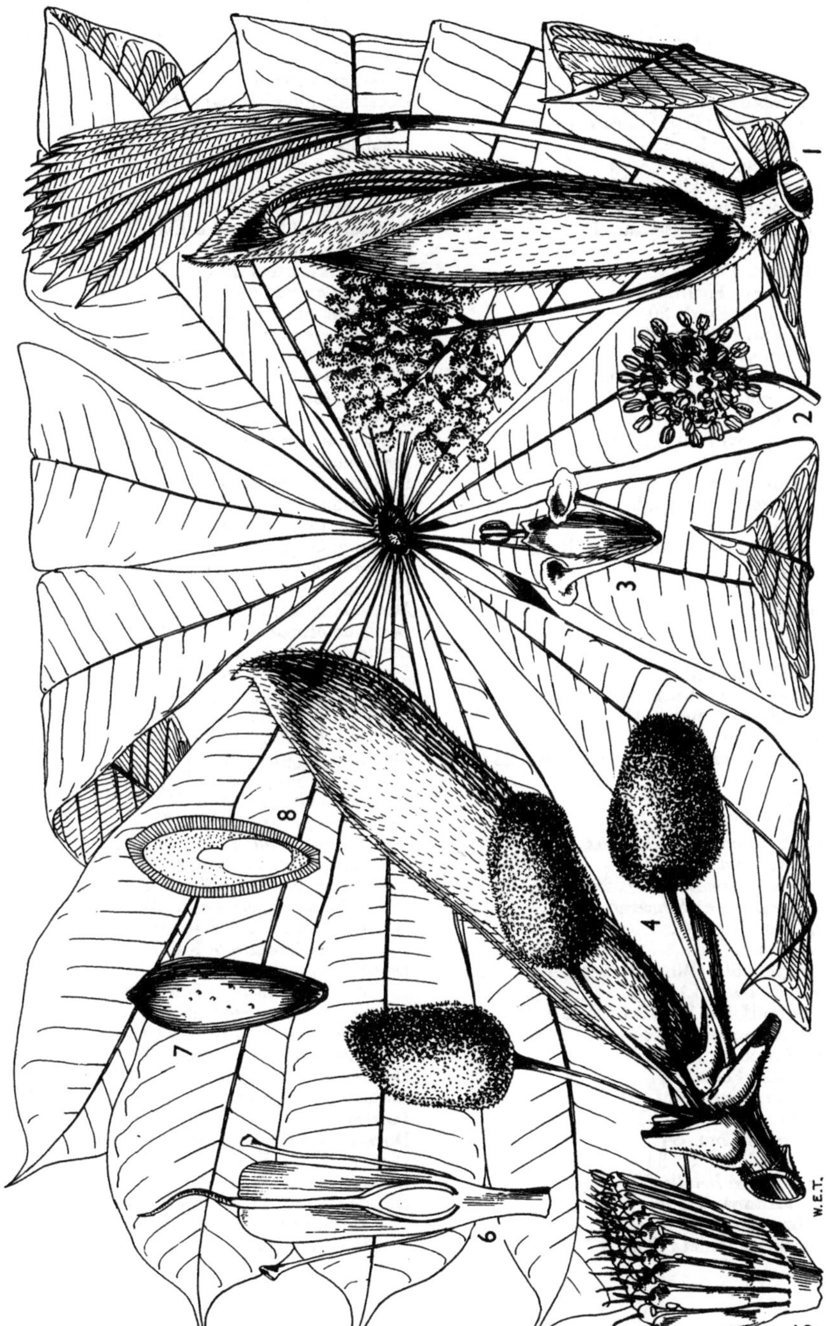

FIG. 23. *MUSANGA CECROPIOIDES* — **1**, male shoot; **2**, head of male flowers; **3**, male flower; **4**, female flower; **5**, portion of female inflorescence; **6**, section of female flower; **7**, fruit; **8**, section of same. Drawn by W.E. Trevithick. Reproduced from Flora of West Tropical Africa.

INDEX TO MORACEAE

GEOGRAPHICAL DIVISIONS OF THE FLORA

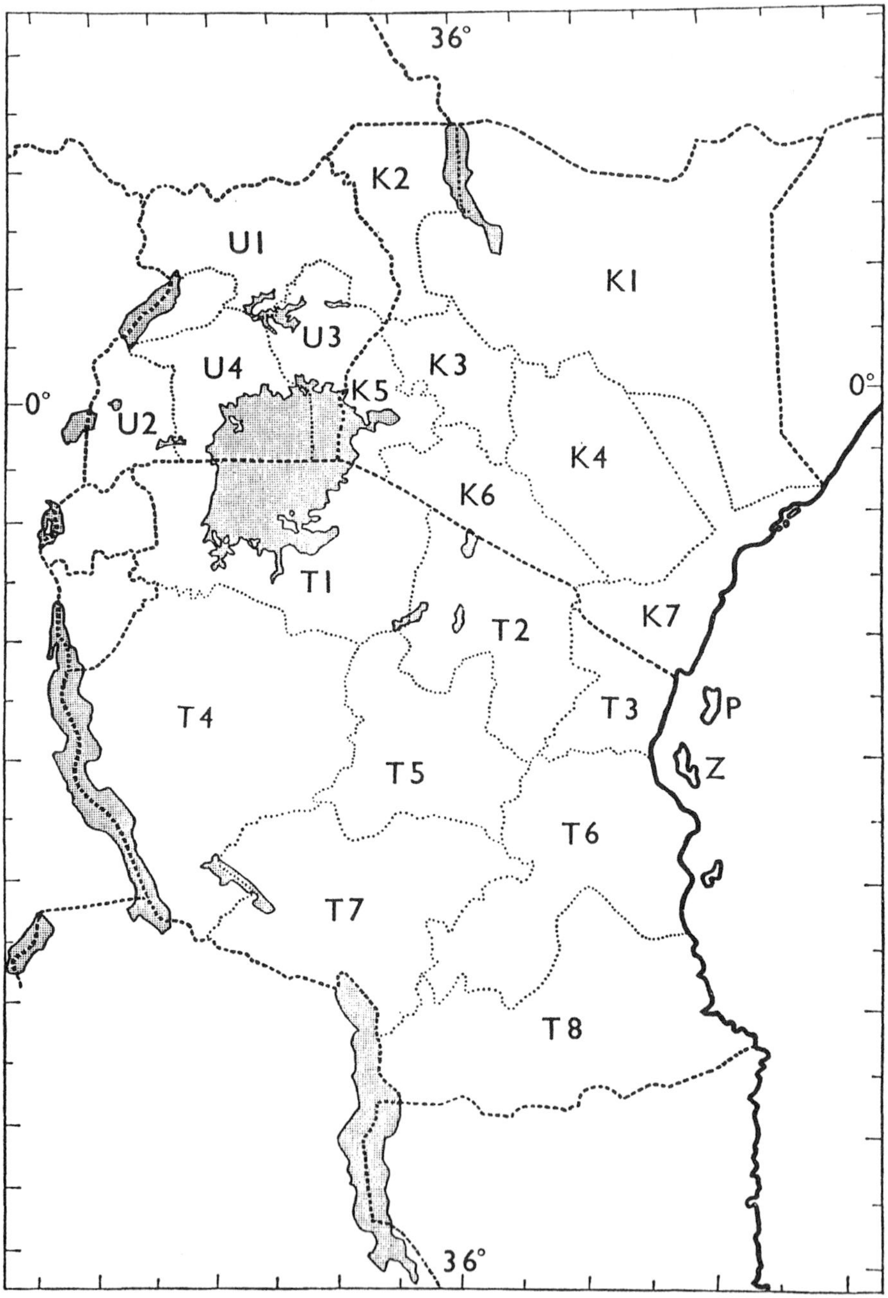

For Product Safety Concerns and Information please contact our EU representative GPSR@taylorandfrancis.com Taylor & Francis Verlag GmbH, Kaufingerstraße 24, 80331 München, Germany

Batch number: 10725827

Printed by Printforce, the Netherlands